INSTRUMENTALITY

INSTRUMENTALITY

INSTRUMENTALITY

On Technical Objects
and Orientations
in the Later Middle Ages

J. ALLAN MITCHELL

UNIVERSITY OF MINNESOTA PRESS
MINNEAPOLIS
LONDON

The University of Minnesota Press gratefully acknowledges the generous assistance provided for the publication of this book by the University of Victoria's Book and Creative Works Subvention Fund.

Portions of chapter 1 are adapted from "Transmedial Technics in Chaucer's *Treatise on the Astrolabe*: Translation, Instrumentation, and Scientific Imagination," *Studies in the Age of Chaucer* 40 (2018): 1–41. Chapters 2 and 3 adopt selected ideas set out in "John Gower Illustrated: The Archer Images, Astronomical Science, and Poetic Identity," *Journal of Medieval and Early Modern Studies* 53, no. 2 (2023): 287–321.

Published by the University of Minnesota Press
111 Third Avenue South, Suite 290
Minneapolis, MN 55401–2520
http://www.upress.umn.edu

ISBN 978-1-5179-1738-8 (hc)
ISBN 978-1-5179-1739-5 (pb)

A Cataloging-in-Publication record for this book is available from the Library of Congress.

Printed in the United States of America on acid-free paper

The University of Minnesota is an equal-opportunity educator and employer.

UMP BmB 2024

CONTENTS

Preface and Acknowledgments

vii

Introduction

1

ONE

INTELLIGENT OBJECTS

19

Planispheric Astrolabe—Ornament and *Organopoiia*—
Multilingual Scientific Cultures—Multiscalar Models—Plane
and Spherical Geometry—Technoscientific Hybrids—
Fictus and Technical Figures—Womb, Mother, Spider,
Horse—Hylomorphic and Epigenetic Change—Intensification
and Extension of Sensation and Cognition

TWO

GRAPHIC INTERFACES

43

Flat Media—Point, Line, Curve—Documents as Devices—
Diagrammatic Figures and Functions—Graphs, Maps,
Games, and Other Inscribed Spaces of Action and
Intellection—Mathematics and the Chord Diagram—Model
Dependence and Metaphorical Transfer—Imagined Archery—
Achieving *Katascopos*

THREE
LEARNING DEVICES, OR INSTRUMENTS OF
LANGUAGE AND LITERATURE
69

Disciplines and *Divisio Scientiae*—Aristotelian *Instrumenta Philosophiae*—Logos in the *Organon*—Grammar, Rhetoric, Poetics—*Possessio* and *Usus*—Subalternation of the Sciences—Stories as Philosophy's Instruments—*Grammaticus, Criticus, Literatus*—Science of Useful Devices—Mechanical Arts and the Invention of Literature

CONCLUSION
TOWARD A CRITICAL INSTRUMENTALITY
91

Critiquing Instruments—Heideggerian Equipmentality—Simondon's Technical Objects and Stiegler's Instrumental Condition—Technocratic Modernity and Instrumental Reason—Academic Working Conditions—Freire's Instruments of Critical Discovery, and Other Available Models—Critical Instrumentality, or Retooling the Arts

Notes
101

Index
137

Preface and Acknowledgments

Instrumentality opens a neglected area of thought for critical reflection and renewal, addressing the pragmatic dimensions of multiple media and cognitive technologies that preoccupied those in the later European Middle Ages. The book offers a reappraisal of principles of utility immanent in such things as time-keeping gadgets, geometrical diagrams, spherical globes and flat maps, mathematical notation and tables, languages and literary tropes, and the very matter and methods of familiar academic divisions of study—logic, rhetoric, and poetics among other arts. Some of these matters have come to shape the familiar apparatuses of contemporary life, equipping minds and bodies still without our perhaps noticing; others will seem more remote. Yet, considered from a historical vantage, various technical subjects and objects disclose long-standing and habitual attachments to a primary instrumentality, a term that, for analytical convenience, can be glossed at the outset as a thing's capacity for work in the world.

The many and diverse modalities of premodern *instrumenta* encompassed a wide array of working mechanisms and media, including things like physical books and philosophical models. Instrumentality pervaded so many areas of medieval consciousness that it is liable to surprise modern readers for whom analogous concepts have developed more restricted meanings. An early and expansive sense of the term is worth recapturing because modern attitudes toward "instrumentalism" have in fact generated narrow and ahistorical views. Now the term is more likely to conjure the ruthless metrics and relentless efficiencies of technoscience under advanced capitalism. Reasons for the stigma are compelling in an emergent technocracy, but the critique has overextended itself to the point of sequestering instrumentality from anything of positive social, political, aesthetic, or intellectual value. Only recently has the term become so anathema, connoting bloodless technicity and transactionalism. We do well to recall a different valence still present in affectionate talk of, for example, musical instruments. For several previous centuries, *instrumentum* was a prominent and productive term of art for describing a host of practical potentialities within philosophy, medicine, law, ethics, rhetoric, literature, and more. Nor will the sense of the instrumental explored in the following chapters be conceived in monocultural terms as a Western notion, for it was a mobile and mutable idea that had long circulated around the Mediterranean and Middle East. The "instrument" is a cosmopolitan concept.

Instrumentality consequently attempts to pry loose a subtle, adaptable, itinerant, and generative concept from limiting contemporary frameworks, and so asks what would happen if technical objects and operations were permitted to affiliate again with inventive, imaginative, pleasurable, or otherwise life-giving pursuits. And in this respect the book amounts to a study of enduring staples of humanistic inquiry. That invocations of an instrumental function may sound somewhat disenchanting to a twenty-first-century ear is reason enough to become reacquainted with a forgotten history of the "tools of life" (Cicero's *instrumenta vitae*). An abiding concern with vital pragmatic forces and effects runs through medieval discourses on ethics and rhetoric, poetry and politics, and the mechanical and liberal arts, wherein what they share, in short, is a commitment to the efforts of practical reasoning. Medieval writers often leave the strong impression that instrumentality is an essential property of *any* productive human endeavor.

It has taken the writing of the book to identify the likely gains of enlarging our sense of instrumentality, and many helped me realize the aims of the larger inquiry. I benefited from conversations with colleagues and friends over recent years and owe special thanks to Iain Higgins, Stephen Ross, Myra Seaman, Lara Farina, Jeffrey Cohen, Lisa Cooper, Anke Bernau, Wan-Chuan Kao, Karl Steel, Matthew Goldie, Sarah Salih, Valerie Allen, Amanda Gerber, Jenna Mead, Seb Falk, Georgiana Donavin, Eve Salisbury, Joyce Coleman, David Carlson, Bob Yeager, and Michael Cornett. Anonymous readers provided generous feedback at different stages, and I appreciate the particular suggestions of David Parisi. I am ever grateful to student interlocutors in graduate seminars on medieval literature and scientific cultures, and for the research assistance of Colin Keohane, Erin Donoghue-Brooke, and Elena Menendez. The Social Sciences and Humanities Research Council of Canada (SSHRC) provided generous funding over five years to carry out the research on which the book is based.

INTRODUCTION

Scientific investigation has long consisted in the search for adequate instrumentation, a working apparatus capable of capturing, converting, displaying, and distributing relevant information. It seeks, in short, eloquent devices. They would constitute a range of things that express and extend knowledge, composed out of an ensemble of practical texts, techniques, and related technologies that make any serious inquiry possible. Medieval practitioners, for their part, recognized the resourcefulness of articulate technical objects, crediting them with an intelligence and productive capacity as operative models. The conceit was occasionally given exuberant expression, as for instance in the spirited prologue to a popular Latin treatise on practical geography and astronomy called *Tractatus de sphera solida* (dated 1303):

> The root and basis of all astronomical theory, and also its immense prolixity and inexhaustible depth of subtlety, started from things that are observed with appropriate instruments. It is agreed among all authorities on this subject that without instruments there would be no way of discovering the motions of the celestial circles and celestial bodies. There is, however, a great number of these instruments. But in this there seems to be general agreement: that they all imitate the motion of the heavens and are made in likeness of it as if after the original [ad exemplar].[1]

John of Harlebeke, to whom the treatise is sometimes attributed, enlisted himself in the tradition as a keen practitioner and propagator of useful knowledge. He was also an avid admirer of instruments. He goes on here to describe the construction of the *sphera solida:* a three-dimensional globe inscribed with celestial coordinates and selected phenomena (poles, tropics, equator, colures, and zodiac constellations), suspended with chords in a cupped base. It is assuredly "easy enough to construct, handsome and delightful in appearance" (*satis facile et visui iocundum et delectabile*), and the "root and model" (*radix et exemplar*) of many other functional devices. Armillary and solid sphere are then mentioned alongside quadrant, cylinder, torquetum, and shadow-instrument, situating the treatise in a long line of making and handling ingenious and attractive technical objects. So many clever mechanisms, transmuting matter and motion into legible information, lend themselves to continued textual production and elevated praise. Empirical and rhetorical properties operate together in such models. John's work would itself go on to become something of a *radix et exemplar,* judging by its broad dissemination.[2]

That animated preamble exemplifies an orientation toward astrophysical models that can be detected in subsequent writings, a variety of works demonstrating the reciprocal relation between technical devices, discourses, and disciplines. Technology shades over into a sort of technography in which the history of scientific devices tends to be enveloped. For instance, treatises on the equatorium from a couple decades later would applaud the competency and dignity of instrumental objects without which the science of the stars could not proceed. The prologue of John of Lignières, *Quia nobilissima* (ca. 1320), states: "Since the very noble science of astronomy cannot be well understood without the appropriate instruments, it was therefore necessary to construct instruments in this field. And so the ancients constructed many diverse instruments, such as the astrolabe and the saphea, with which much can be known about both time and motion."[3] Nodding to earlier practitioners, John takes pains to orient the work within the longer trajectory of instrument-making, singling out the proficiencies not just of the astrolabe and saphea, but also the armillary, solid sphere, torquetum, cylinder, and quadrant. Providing historical conditions for ongoing scientific developments, such devices also construct the contexts in which an encomiastic rhetoric of instrumentality is generated and sustained over time and space. The text begins to tell a story in which instruments are celebrated and mobilized as functional and communicative objects. Assuming a comparable attitude to matters of practical science near the end of the same century, Geoffrey Chaucer's *Treatise on the Astrolabe* (1391) likewise remarks on the intellectual authority and agency of the planispheric astrolabe, suggesting that the device may well know more than he can even tell: "truste wel that alle the conclusions that han be founde or ellys possibly might be founde in so noble an instrument as in an astrelabie ben unknowe parfitly to eny mortal man in this regioun, as I suppose."[4] Another talented technical object with a distinguished history, the astrolabe, as we will see in detail later, conveys a palpable sense of how knowledge is transmitted in and through capable instruments. Knowers come to rely on the objectification and intensification of knowledge in a captivating technical apparatus, leading at times to exciting and unexpected "conclusions."

How do technical objects come to be perceived as such fascinating *knowing* things, operating to capture and convey information about the ambient environment? To what degree might a sense of instrumentality be extrapolated across a spectrum of knowledge practices, technologies, attitudes, and institutions, describing a broader epistemic orientation? And what counts as an instrumental apparatus anyway?

Answering such large questions inevitably involves successive redefinitions and sampling from a constellated array of medieval technical devices, discourses, and disciplines. We can begin with the proposition that an operative instrument is one that works as an intermediary and expressive interface, supplying the enabling conditions for human speculative and practical understanding. I have chosen a small selection of working instruments that, given their complementary relations, are especially suggestive of this idea, registering what I take to be a capacious premodern sense of the affordances of multiple media technologies (e.g., we will observe how instructional texts describe instruments that in turn inform written texts and visual illustrations). The basic thought is that competent technical objects and operations stand in vicarious relation to subjects engaged in knowledge production and mobilization, showing just how much human intellection and action are, in profound respects, *attributes* of a given *instrumentarium*. Francis Bacon would identify the same epistemic preconditions of the natural sciences a couple centuries later in his *Novum organum* (1620), or "The New Instrument." His second aphorism begins: "Neither the naked hand nor the understanding left to itself can effect much. By instruments and auxiliaries the work is done."[5]

That premodern instruments were not simply serviceable but also intellectually engaging, inventive, imaginative, pleasurable, and historically and culturally consequential is well captured by the descriptive accounts so far, both in Latin and in the vernacular. What they can begin to adumbrate is a notion of technical devices as instrumental in a wider sense than that term is now often taken to signify. Insofar as spheres, equatoria, and astrolabes equip practitioners with tools to measure and model phenomena, they open up parts of the world to novel apprehensions, mediation, and interpretive processes. All devices have their peculiar histories, but in a potent sense, they are also *telling* objects. The ones I have mentioned so far encode and express temporal and spatial changes within the environment, even simulating current and future terrestrial and celestial arrangements. They are articulate objects. And some multipurpose technical instruments were marvels to behold ("handsome and delightful in appearance" or, repeatedly, "noble"), appearing to be uncanny, even charismatic things of underappreciated beauty and untapped potential. What sort of intelligence could such things possess? How do they render phenomena visible, legible, navigable, or otherwise usable? To what extent are they seen as capable of extending and intensifying sentience and skill? And do they not often produce forms of recreation and enjoyment?

Instrumentality may well be the name people give to productive, co-constitutive relations between subject and object, mind and matter, human and other-than-human agencies in a given technical apparatus. And they are manifold. "Medieval instrumentality" quickly gathers within its bounds more than one kind of affordant entity. We can already discern that instruments belong within technical ensembles that integrate companion texts, theories, histories, habits of mind, sensory feedback, and receptive physical materials within situated environments. While my opening examples are of paraphernalia associated with an area of natural philosophy, Bacon's aphorism indicates there are others sorts with which to grapple in the history of technical objects and instrumental orientations. These include members of the body (*manus nuda*) and the mind (*intellectus*), but also much else besides, including images, methods, documents, disciplines, and schools. The historical scope of the term will show how far instrumentality stretches concretely and conceptually, applying to persons and impersonal things within what we might call emerging premodern "technocultures."

In English, the earliest recorded use of "instrument" in a technical and scientific sense appears in Chaucer's *Treatise on the Astrolabe*, referring to an altimetric device he describes and whose functions he demonstrates in a step-by-step manner. "The firste partie of this tretys shal reherse the figures and the membres of thyn Astrolabie, bicause that thou shalte have the gretter knowing of thyne oune instrument" (662). Deriving from the Latin noun *instrumentum*, the word came from the verb *instruere* (to build, arrange, prepare, inform, or instruct) and had already applied in varying contexts to tools, musical instruments, legal documents, catalyzing agencies, and bodily members or whole persons.[6] It is a term that could easily cross the thresholds between counting, writing, arguing, making, imagining, and performing, delimiting one or another site of creative expenditure. So many senses are possible when dealing with such a useful lexeme, a word adept as a finely calibrated instrument for mediating, modeling, and mobilizing knowledge. What seems common among early senses is, quite simply, a determined capacity for transformative work. For this reason, instrumentality can encompass a rich variety of human endeavor—calculation, literary composition, cultural criticism, and historiography among them—even coming to describe the ideal course of a liberal education. That the term is so flexible and expansive makes writing about such phenomena challenging and engrossing, rewarding a sustained analysis of the many technical objects and orientations that align under the general rubric. An instrument, at some fundamental level, seems to make and multiply meaning. Isidore of Seville's *Etymologies* provides a neat epitome: "An *instrumentum* is what we use to make something, as a knife, a reed-pen, an axe." Each of those items can result in a particular construct (*instructum*), "such as a staff, a codex, a table." And each of these has its own use (*usus*), "that use to which we put the thing that is made, such as learning on a staff, reading in a codex, gaming on a table."[7] A cascading series of causes and effects rapidly flows from the way one instrument constructs other things that are usable, becoming instrumental in their turn within a longer chain of causality and expressivity, as if generating increased throughput in a given system. In his book, Isidore dedicates a vast number of entries to the many and varied *insrumenta* of music, law, farming, gardening, horse riding, rhetoric, medicine, metalwork, war, carpentry, and cloth-making, eventually applying the term to the work of antiquaries and writers like himself. Just so, a "scribe's instruments are the reed-pen and the quill."[8] As Isidore is reporting in this usage, writing is instrumental in being a propagative mode from which derive other generative acts—a pen is for making a book, a book for reading, and reading for various applications—all of which is instantiated in his own *Etymologies.*

Other products of the pen are just as impressive and potentially enriching for being so instrumental, and the present book is preoccupied with principles of utility that inhere in some foundational knowledge practices. Here too, instrumentality has multiple meanings, not least because we are engaging the very components and techniques of sense-making. As we will see, Aristotle's *Organon* (Greek for "instrument") looms large in the intellectual history of instrumentality that would eventually expand beyond logical disputation to include other modes of communication and higher learning. Initially, as Isidore himself observes, a philosopher investigating first principles of metaphysics has recourse to elementary *instrumenta* set forth in speech and writing, with the aid of tools of predication or categorization, which is the initial lesson of Aristotle's *Categories.*[9] Those are, so to speak, the principal tools of the trade. By the twelfth century, rhetoric and sometimes poetics were treated as auxiliary instruments with just as much share in the development of the arts, both liberal and mechanical. Hugh of St. Victor spoke of the arts curriculum as a whole as "instruments of all philosophy" (*totius philosophiae instrumenta sunt*),[10] using language that would reverberate through the works of such scholars as Thierry of Chartres, John of Salisbury, and Robert Kilwardby. Nor did the term demarcate a single discursive context or academic curriculum. Sometimes the matter was clerical. Monastic meditation relied on rhetorical and mnemonic *instrumenta,* religious devices useful for lifting the spirit, as for instance in the way the Benedictine Rule expressly speaks of the "Instruments of Good Works" (*instrumenta bonorum operum*), or practical precepts for living and working together in community by plying the craft in a spiritual workshop.[11] Other examples were didactic and pertained to a range of general and specialized subjects. Tasking readers to make culinary recipes, play games, craft objects, hunt game, or plant and graft trees, how-to texts were diverse and increasingly abundant.[12] Still other instrumental works are to be identified within the realm of imaginative literature, devising powerful engines of thought and feeling in the form of fictional narratives.[13] Literary art was ideally disposed to produce a complex mix of use and enjoyment—evoking a premodern sense of their instrumentality.[14] That written compositions could become such important instruments of intellectual inquiry, personal development, or collective endeavor lies at the center of this study of how varieties of technical sophistication work in the world. They were treated as something like a productive or transformative organ of sense and signification. Though spiritual texts and a great deal of pragmatic material will not receive all the attention they deserve in this book, we can pause to notice the nearly ubiquitous terminology. Here it becomes clear enough how medieval thinkers sought to accommodate *instrumenta* to so many recreational, devotional, and educational ends.

The sentiment and semantic ranges extend further still to encompass working parts and processes of the organic body, as evidenced in several antique and medieval descriptions of the human perceptual apparatus. *Instrumentum* has a sense that stretches to include the physical sensorium. The idea was given expression in different ways via Cicero's and Seneca's "tools or equipment of life" (*instrumenta vel ornamenta vitae*) and Boethius's "apparatuses of the senses" (*instrumenta sensuum*), on down to Robert Grosseteste's "motive powers and instruments of the body" (*virtutes motive corporis et instrumenta*).[15] Those ways of viewing the mobile and sensate body often look to specify with some precision the uses and effects of anatomical components, the practical relations and expenditures of which provide for human life. The notion was venerable. *Organ* and *organism* come from Greek via Latin *organon*, which was synonymous with *instrumentum*.[16] And the idea was durable enough. Medieval encyclopedic sources, such as Bartholomaeus Anglicus's thirteenth-century compendium *On the Properties of Things*, elaborated on the body as "þe propre instrument of þe resonabil soule in his workes of kinde and of wille."[17] In his Aristotelian understanding, the hand is given pride of place and demonstrates a human dependence on, and internalization of, external equipment. For Aristotle, the mind is, like a physical hand, an "instrument of instrument" in being so versatile and open to supplementation or augmentation. The comparison evokes a primary human instrumentality.[18] So Bartholomaeus writes, "þe hond is a greet help and ormanent of þe body, and propir principal instrument of touchinge."[19] The word *ornament* itself could mean both adornment and equipment (mingling art and nature), recalling again the boundary-crossing, mediating functions of instrumentality.[20]

Early technical objects and discourses consequently establish unexpected connections between persons and impersonal things, corporal and incorporeal matters, orienting interior and exterior organs around ready-to-hand instruments. All can be characterized by the way something auxiliary to the human form is made to fit and function prosthetically, augmenting and enhancing the natural body. Some were the products of apparently wild imaginings, as exemplified by Roger Bacon's catalogue of artfully contrived instruments, like swift self-propelled watercraft piloted by a single man (*instrumenta navigandi*), or aircraft that flap and fly like birds (*instrumenta volandi*), or small contraptions for elevating and lowering heavy bodies (*instrumentum altitudinis*), the last of which he thought might come in useful for escaping prison, of all things.[21] These might have been no more than promising tricks or gimmicks, while others are seemingly more essential or established members of the body. They include things as small and unassuming as the fingers of the hand. Glossing Aristotle again, Bartholomaeus writes: "Aristotle seiþ þat þe hand is not on instrument but manye."[22] From early on, the palm, fingers, and joints served not just manual tasks but also memory, communication, and computation.[23] All instill the digits with a capacity not just to retain information but also to visualize and implement abstract modes of cognition. A growing appetite for vernacular "handbooks" and "manuals" (from *manus*, "hand") shows how books emerged as another external apparatus that recomposes remote ideas for prehensile beings, which is one way to look at the function of electronic devices today. Articulating across seemingly separate domains, the human finger is still central to technical practices and systems, having become a founding figure of the digital (from *digitus*, "finger"). Reading and writing are, in a sense, how hands work. *Instrumentality* will be concerned with the interdigitation of the mind and body, which is not just a metaphor.

As this sketch suggests, persistent if not always self-evident aspects of instrumentality warrant further investigation and conceptual renewal because the notion embraces so many objects and operations that equip the living. And herein lies the main summons of the book: to resist for the moment reductive senses of the instrumental as extraneous rather than essential, extractive rather than constructive and transformational, constraining rather than expansive and capable of initiating new imaginings. There is as yet no history of the topic that cuts across what have come to seem such distinct domains and disciplines. *Instrumentality* is a book that explores an indispensable crossover concept that travels the circuit between speculative and applied knowledge, integrating diverse subjects and objects within vital working assemblages. Premodern instrumentality shows how individual agency and intellection is defined and distributed, enabling situated and skilled knowledge practices still recognizable today.

But why take up the matter of instrumentality now? The key term at the center of this book has undergone an abrupt pejoration in modern theoretical discourse, where it is likely to summon the dead hand of mechanical influence, reductive quantitative measures, dehumanizing routines, and eventual technoscientific domination, offending the ethos and enlightened consciousness of the modern intellectual. It is not a word like "craft" or "artisan," whose senses have come to exert a nostalgic pull, even if the competencies denoted by them require so many tools and technical orientations.[24] Critiques of "instrumental rationality" obtained particular gravity in the social and political theory of Max Weber and the Frankfurt School, reverberating ever since among those committed to the singularity or transcendence of artistic or cultural phenomena. "Instrument-knowledge" has come to seem at odds with poetry at least since Matthew Arnold.[25] The salient tension is older still if we believe the founding gesture of ancient thought was to separate *technē* from *epistēmē*, distinguishing technical skill from language, reason, and knowledge.[26]

I will return in the final pages of the book to survey major objections, by which time we will have a clearer view of how much *criticism* relies on technics and analytics that tend to be disavowed in their implementation, ultimately running the risk of distancing the skeptics from the conditions of their intellectual labor. Anti-instrumentality is not really a viable option. Readers are invited to turn to the conclusion now if they would rather not wait for a more extended and relatively polemical engagement with the critics. My argument eventuates in ideas that will have purchase on present-day debates about the value of intellectual labor in relation to the rhetoric of instrumentality rather than despite it. Ultimately, the present book challenges what has become a dominant paradigm of literary humanism with its overestimation of individual talent alongside a concomitant underestimation of the requisite tools of thought and conditions of collective existence. *Instrumentality,* if the reader will permit the wordplay, stands in tension with another mentality that prizes the singular human intelligence or retiring "life of the mind." As William Clark shows in his account of the development of the modern research university, scholars acquire authority "with or even through an armory of little tools—catalogues, charts, tables (of paper), reports, questionnaires, dossiers, and so on."[27] Academics are also subject to so many administrative devices and ministerial documents that rationalize academic labor, many of which stem from medieval and early modern practices; and humanist scholars themselves create, and continue to innovate on, the devices, documents, social norms, and institutional operations that structure academic work. That partly motivates the present reckoning with instrumentality. The liberal arts disciplines have, it is clear, long required a suitable apparatus to furnish the conditions under which subjects can be studied and taught. We will observe the degree to which technical objects and orientations are bound up with innovative, expressive, and imaginative artifacts, inscriptive or visual communicative devices, and any rhetorical and literary media, enabling pretty well all learning and teaching. What the foregoing examples begin to suggest already is that instruments are ever timely objects, exerting force, catalyzing energy, extending sense, propagating knowledge, or otherwise configuring historical possibilities. Instruments, in short, hold reserves of causal agency or productivity. Many still litter our everyday lives, including maps, clocks, books, legal fictions, and disciplinary formations, as, in sum, assistive technologies. Such objects sustain social and intellectual life. They equip subjects with fine-tuned interpretive devices. They are companions to large fields of inquiry. And a diverse tool set is now more urgent than ever if we are to address public health crises, climate change, racial capitalism, and neoliberal globalization, all practical problems wanting cross-disciplinary solutions. Let us contemplate then a critical instrumentality, where the emphasis lies on learning to use all available tools and techniques to enact positive social change.

The full magnitude of the argument will come into sharper focus after considering a set of case studies in the chapters below, but I want to say just a little more about how my attempt to recover the lineaments of medieval instrumentality fits with some recent scholarly developments. Notwithstanding leeriness in some quarters, surely now is an auspicious time to take stock of instrumentalizing matters from a historical vantage informed by recent theories of media and technology. The ground has been cleared to revisit the operative modes and meanings of premodern devices, building on work in several related subfields. We can tell a longer and better story about technical objects and orientations thanks to generations of historians of science and technology, curators, collectors, and critics. They have taught us that a restricted definition of "scientific instrument" is hardly sustainable in a past cultural moment when the term could comprehend so many diverse phenomena.[28] Literary and cultural historians have also done much of late to reframe medieval technical ingenuity in relation to genres, media technologies, information systems, cartographic imaging, and manuscript production, and my debts to a large and growing body of work will be evident. One of the marks of medieval literature is what scholars have called "instrumental forms."[29] In broad terms, the present book should complement a recent turn to considerations of knowledge creation, organization, and mobilization, and especially to the functioning of media and technology in the Middle Ages.

For the effects of technical mediation to register more decisively, historical scholars would benefit from a rebooted critical methodology to restore fundamental competencies to technical objects in what have too often been considered non-overlapping domains. An important impetus can already be found in digital humanities, media archaeology, and platform studies that pay close attention to technical methods, media substrates, virtual interfaces, and machine processes.[30] A forensics of digital mechanisms and systems could be just as productive in the study of medieval media technology. Another stimulus for the book is the spate of nondualist philosophies and feminist science and technology studies on the agency of physical objects and systems, stressing human dependency on assistive tech among other artificial resources, embodied most keenly in the prosthetic.[31] "Living bodies," observes Elizabeth Grosz in her elegant formulation, "tend towards prostheses: they acquire and utilize supplementary objects through a kind of incorporation that enables them to function as if they were bodily organs."[32] And Bernard Stiegler theorizes: "A pros-thesis is what is placed in front, that is, what is outside. . . . The being of humankind is to be outside itself." For such thinkers, human ontology emerges out of an "instrumental condition," combining the capacities of *technē* with those of *logos*.[33] Karen Barad's account of the scientific apparatus is equally enriching: "The notion of an apparatus is not premised on inherent divisions between the social and the scientific, the human and the nonhuman, nature and culture. Apparatuses are the practices through which these divisions are constituted."[34] Some of them are highly abstract, mental supports (already implied in Stiegler's *logos*). I have also learned from theoretical treatments of instruments of thought, notably Brian Rotman's work on mathematical and alphabetical subjects and Catarina Dutilh Novaes's idea of formal logic as "cognitive technology."[35] Mechanical, mathematical, logical, and literary formalisms all reveal what abstract inscriptions can accomplish, modeling worlds rather than simply describing a given one. That is one of the points of recent work on the parallels between mathematical and literary forms, something we can probably find in earlier epochs if we care to look.[36] My sense of the possibilities involved in quantitative thinking is likewise indebted to Steven Connor's energetic defense of numbers against a contemporary "anti-numerical prejudice."[37] For they too are critical instruments that convey multiple values.

Much of the relevant scholarly work has been focused on new tools and infrastructures (if not futuristic cyborgs and technological dystopias), but this book will contribute a historical point of view. What are the advantages? One is that a notion of medieval instrumentality should thwart an impulse to situate early technocultures in a line of inevitable succession that culminates in the worst excesses of the present. Even media archaeology conjures metaphors of time that are stratifying, developmental, and effectively supersessionist, seeing in the historical record a continuous layering over and outmoding of the past.[38] Some critics note in the emergence of the "media concept" the very occlusion of the "medieval," and are pushing back against the resulting periodizing and presentist paradigms.[39] It is surely more productive to see in the predigital past some of the operative technics, analytics, and metaphorics that still structure human experience, and that go to show how moments we think are "over and done" continue to have momentum *now*.

Another benefit of focusing on earlier technical objects and practices is to resituate them within historical and cultural horizons where they augmented and affirmed human desires and disciplines, and so defamiliarize what it means to mediate or instrumentalize *then*. For, premodern *instrumenta* typically cannot sustain future distinctions between the scientific and nonscientific, nature and culture, bodily members and impersonal mechanisms. To that extent, they refocus our sense of what is possible. Just so, a recent essay collection devoted to premodern prosthetics pushes against the sort of futurism that has defined many other studies (e.g., celebrating the bionic or virtual) and shows how "medieval and early modern prosthesis offers to speak to—and maybe even re-assemble—our present-day discourse on this subject."[40] Part of the force of historicizing such notions as "media" and "prosthesis" and "instrument" comes from the way their antecedents seem perhaps alien, opaque, or borderline. The "alleged marginality" of old technical devices, as Thomas Hankins and Robert Silverman observe, is useful to the extent that margins are "surfaces of contact and connection between and among different themes and entities."[41] We will take up examples of medieval imaginative literature that incorporates technical and scientific matters (just as some medieval scientific descriptions take the form of poetry and include imaginative, speculative, or fictional devices). Such cases model the interrelations of the technical and aesthetic that elude later periods that have rather tended to cultivate and enforce increasingly attenuated disciplinary norms and distinctions.

Instrumentality composes a history of a useful concept that has for centuries powered and sustained practices we can recognize as profoundly interdisciplinary. What makes instrumentality of special interest, then as now, is its capacity to cultivate a habit of thinking relationally, synthetically, inclusively, and expansively, resisting the compartmentalization of disciplines within respective academic silos. It turns out I am not just collecting curious and companionable examples, but also exploring prospects for a renewed liberal education. In this respect, the book is a work of cultural and intellectual history ever on the verge of a general theory, and I hope readers enjoy the scope and speculative opportunities the work affords.

Nor were technical subjects so monocultural as they have become today in professional scientific fields, and that imparts important lessons as well. Any detailed study of the history of science and technology will be alive to the profound and complex intimacies of global cultures around the medieval Mediterranean. Medieval scientific practices and instruments could forge intimate affiliations between disparate times, places, polities, and matters, and were the result of dynamic international text networks and transnational exchanges. This is to distinguish premodern practices from scientific culture today, which Michael D. Gordon argues is the result of moving from a plural and polyglot "Scientific Babel" to one that is now narrowly Anglophone.[42] STEM (science, technology, engineering, and mathematics) concentrations are known for limiting scope and outcomes in this way, with funding disparities and institutional mandates exacerbating the problem. But, even when technical objects and orientations appear to serve hyperspecialized or parochial purposes, they remain entangled in lived practices that involve multiple agents, media, materials, and networks.[43] Medieval sciences studied in this book are the more conspicuous for participating in interdisciplinary and international milieus, mobilizing transcultural knowledge practices—adoption of Hindu-Arabic numerals into Latin and English (sometimes written out right to left in imitation of Arabic sources), successive improvements on traditional instruments and techniques (embracing technical developments from Baghdad and Alexandria), various and overlapping translation projects of Muslim, Jewish, and Christian scholars working together—and could, at times, have been seen as propagating ecumenical outlooks. As will soon become clear in this book, a set of transnational exchanges, text networks, and translations supported educated inquiry and instrumentality. International knowledge practices were critical in sourcing and sustaining "cosmopolitan language systems," producing a kind of "travelling wisdom."[44] Not that it was all frictionless global collaboration; but discrete processes of knowledge transfer and mutual multilingual discovery yielded more than traces of cultural interchange and enlarged hemispheric horizons, often quite literally so, as we will see. My aim is to examine verbal inscriptions, visual media, and mechanisms as interlocking parts of a historical assemblage at the juncture of early technoscience, all to better make out the main lines and effects at a distance, some of which endure without being widely recognized.

Instrumentality consequently examines practices and principles that were as profoundly international as they were interdisciplinary in scope. Granted, my limited expertise means that examples will skew toward a subset of cultural artifacts and practices: historical sources discussed in the following chapters will tend to circulate around the European Middle Ages, with English and Latin literature furnishing contingent touchstones. Evidence nonetheless comes from a time and place of immensely fertile cross-cultural exchange and technical and intellectual innovation, spanning roughly 1100–1500, and my examples are chosen for the way they open onto broader transcultural horizons and global histories, articulating among roving concepts and practices. Instrumentality is hardly a Western medieval invention; its roots lie elsewhere. John of Lignières, mentioned at the outset, was a prolific Parisian scholar who disseminated astronomical data first compiled in Castilian. Chaucer composed an English translation of a Latin treatise originally deriving from the work of the Jewish-Arab astronomer Māshā'allāh ibn Atharī. Those sciences were cosmopolitan pursuits. Many other practitioners likewise stood at the crossroads of Hebrew, Arab, Greek, and Latin science. It is for specialists in other fields today to discover whether or how far particular ideas may be relevant within adjacent periods and areas of inquiry, and to take up and experiment with the central intuitions that have come down through the works I study: *that instruments intensify and augment an individual and institutional capacity for work.* My book is consequently as provisional as the models it examines, pursuing a historical inquiry in a pragmatic manner that requires further testing and theorizing. We will see that, as ancient and medieval writers well knew, to model something is to work within the constraints of what a given device can demonstrate, as conceived in a tradition extending back to Ptolemy's ὀργανοποιία (*organopoiia*) and forward to the latest thinking about the hypothetical nature of a given scientific apparatus.

The book is divided into three analytical chapters and a comparatively polemical conclusion, moving progressively from concrete to more conceptual technical objects and orientations affiliated with premodern instrumentality. Subsequent chapters tend to build on information imparted in earlier ones and pursue a series of interlinking paths that can be seen as loosely modeled on the *viae* ("ways") structuring the liberal arts according to the *trivium* (grammar, logic, and rhetoric) and *quadrivium* (arithmetic, geometry, music, and astronomy). Some pathways and intersections may now seem strange, but the itinerary has many points of interest that allow us to apprehend the circuits of thought that joined distinct knowledge practices under the large rubric of instrumentality, or *philosophiae instrumenta.* Specifically, the three chapters advance from examples of (1) three-dimensional measuring devices whose material and mathematical properties are central to their functions, to (2) two-dimensional diagrammatic figures where the emphasis lies on verbal and visual interfaces, to (3) rational and rhetorical *instrumenta* allied to overarching disciplinary norms. The sequence is intended to uncover a series of reciprocating forces among related cognitive operations and curricular developments, though they were not all contained within some set university curriculum; they were rather diffuse and aspirational in attempting to capture the interplay of epistemic objects and trained-up subjects, even when dealing with very rudimentary acts of intelligence.

Instrumentality ultimately hopes to reacquaint readers with vital and congenial attitudes toward technical objects, operations, and orientations that extend beyond the Middle Ages. In a concluding commentary, I spell out some implications for those of us engaged in criticism today, who in some measure carry on the legacy of pre-modern knowledge practices. One hoped-for result of the book is to shake up naive notions of academic detachment associated with a pervasive anti-instrumentality. I press hard against an intellectual aloofness that could never have created the institutional conditions in which scholarly disengagement was possible in the first place. Yet the general disavowal of anything so vulgar as "utility" or "productive activity" within the humanities threatens, I think, to constrain radical thought and action. Those engaged in the study of media and culture across a variety of fields should rather acknowledge and affirm the need for well-equipped action and intellection, whether urging vigorous critique, slow scholarship, or some form of aestheticism, and whether criticism is to be characterized by up-to-the-minute topicality or something rather more arcane. Anticipating my final thoughts, the question has never been whether to employ instruments, but which ones are most conducive to the work undertaken. This book goes some way toward locating historical grounds for what I will call "critical instrumentality," advocating for the indispensability of practical reasoning.

INTELLIGENT OBJECTS

Composed late in the fourteenth century but never completed, Geoffrey Chaucer's *Treatise on the Astrolabe* describes a multipurpose mechanism (all-in-one clock, compass, calculator, star map, and data-storage-and-retrieval device) whose primary reference is the predictable arrangement of empirical phenomena. Two of five projected parts make up the unfinished work: the first details the physical components of a planispheric astrolabe; the second covers over forty possible applications (called "conclusions").[1] Tables with auxiliary data were intended but not included in extant versions. The spare expository prose together with the impulse to diagram and tabulate reflect a deliberate pedagogic strategy, prepared as it was for training up a ten-year-old boy, "Lyte Lowys my sone" (1). Adapting practical matter to the needs of a young amateur, Chaucer embraces an artless style ("full light reules and naked wordes in Englisshe"; 21–22) and employs purposeful redundancy ("my superfluite of wordes"; 36). The instructions progress incrementally with a systematic explication of use cases, successive instances building on previous ones. The first describes how to find the sun's longitude by setting the alidade on the back of the astrolabe to the day of the month and then reading off degrees; the second pursues the altitude of the sun, moon, or another celestial body by suspending the astrolabe from the right thumb and rotating the alidade until sunlight shines through the holes, before noting the degree of altitude; there follows a third technique for finding the time of day or night, and so forth. Chaucer attends throughout to pragmatic means and soluble ends, issuing a stepped series of operations and modest calculations to arrive at repeatable and verifiable results, insisting on procedural clarity and correctness. The astrolabe emerges as a practical instrument, affording efficient modes of information capture, conversion, and display, based in sound observational methods.

The sheer instrumentality of the written work is bound to be somewhat surprising, if not off-putting, to students of contemporary literary culture, not to mention those familiar with the imaginative works of Chaucer. Chaucer's *Astrolabe* appears to be a no-nonsense training manual, addressed to expedient and utterly banausic ends. Accordingly, to those expecting more in the way of the poet's signature irony and irresolution on display in *The Canterbury Tales* or *Troilus and Criseyde,* the treatise can seem relatively uninspired, plainly pragmatic, frankly expository. Yet this is, if anything, a modern problem. Notably, Chaucer's role as translator and technical writer had received appreciative nods early on. Within a few decades, his *Astrolabe* was circulating broadly, in over thirty manuscript copies, more than for any other work by Chaucer outside of the *Tales*.[2] Chaucer himself anticipates a kind of intergenerational kinship, beyond little Lewis, with "every discret persone that redith or herith this litel tretis" (34–35). But what could explain its broader appeal? "Embarrassed by the apparent popularity of a work that seems so foreign to our modern view of Chaucer's attractions," Simon Horobin observes, "scholars have sought to explain away this situation as an aberration."[3] It is a curious situation that exposes unexamined assumptions about literary and technical disciplines, then as now, and immediately draws us deeper into underrecognized aspects of medieval instrumentality.

We must not underestimate the gravitational pull of scientific instruments in the period; nor should we neglect a set of intellectual, aesthetic, and affective concerns constellating around the astrolabe long before Chaucer. How might readers better recognize the expansive meanings of something so evidently matter-of-fact, mechanical, and instrumental? What would it take to grant the prosy, technical object primacy again? How can we grasp an astrolabe's speculative and pragmatic principles of utility?

There is something sensational and diverting about planispheric models like the astrolabe, limning connections between *use* and broader considerations of art, animacy, cognition, and embodiment. Technical thinking associated with scientific instruments was a rich source of what Edward Grant has called the medieval "scientific imagination."[4] For young Lewis, the whole contraption must have been an enchanting science kit, a gift of neat technical gadgetry complete with swappable plates, globe, plumb line, pen, and instructional manual (Figure 1). Mature practitioners already familiar with practical astronomy also often felt the allure of such inventive models, techniques, and hypotheses associated with the branch of natural philosophy, and Chaucer's description registers his own keen delight in "so noble an instrument" (105). Sometime before him, Roger Bacon's creative faculties had been ignited by the thought of a self-propelled spherical planisphere (*instrumentum sphaericum*) whose whole array is set in motion by the natural course of the heavens.[5] Abelard and Heloise would name their son Astrolabe.

Derek de Solla Price recalls that astronomical instruments were once called *theorics,* which suggests they are compelling speculative or hypothetical things, objects of fascination.[6] Here we begin to detect an attitude toward instruments that had by the twelfth century won them a firm place in philosophical discourse and what became known as *scientia ingeniorum* (the science of devices), about which more will be discussed later in the book. For now, it should be enough to suggest that such things as astrolabes and equatoria belonged to a way of approaching technical objects as more than mere appliances; they are also stimulating models, elegant simulations, and spectacular ornaments. Chaucer devotes himself to a phenomenal epistemic medium at least as ingenious and expressive as any literary composition, indeed "poetic" in the older sense of constructing figures, engaging in a kind of *technopoiesis.*[7] Only, here, we are not speaking of fashionable letters. The poetics of the technical object rather springs from a venerable practice of ὀργανοποιία (*organopoiia*), or instrument-making, going back at least to the famed Alexandrian named Ptolemy, for whom specialized scientific knowledge was to be instantiated in curious devices.[8] It is precisely from a mixed pragmatic-theoretic, model-dependent, object-oriented vantage that the technical object should compel interpreters, especially those in the arts or humanities, to reconsider what it can mean to instrumentalize. One the one hand, our sense of the barefaced technicity of the work poses a useful stumbling block, for Chaucer's *Astrolabe* is sufficiently sincere and serviceable to perform a productive decentering of what has become a model literary

FIGURE 1. Earliest English astrolabe, dated 1326 (British Museum, MLA 1909, 6–17.1; *Epact* 40428). Copyright Trustees of the British Museum.

subject. On the other hand, what comes into focus is the intellective and imaginative modality of an engrossing medium of geometry, geography, and practical astronomy in the astrolabe. As artful as it is operational, the ingenious device captures something of the cosmos it describes and sets in order (recalling that Greek κόσμος can mean both "ornament" and "order"). Aware of the implications, Stephen G. Nichols provocatively suggests that Chaucer's description is "essentially an ekphrasis of a work of art (the astrolabe)."[9]

That *art* is embedded in the instrumentation takes us some way toward perceiving how instruments are constitutive of the liberal and mechanical arts, which is where the present book is headed. How a specific technical object such as the astrolabe communicates across such seemingly incommensurate domains as literature and science serves as the focus of this chapter, before the question broadens too far. Chaucer's *Astrolabe* is preoccupied with a device that is at once mental and material, mimetic and pragmatic, aesthetic and empirical, and somehow animating while remaining inanimate. It is, as we will see, a prosaic work that happens to be especially lucid about the way instrumentation actively embodies and encodes information, externalizing and intensifying knowledge in a variety of modes. A scientific apparatus of this sort propagates perspectives and positional information, enabling access to otherwise remote regions and ideas, while also generating surprising associations and exhibiting particular charms as a crafted objet d'art. Taking the measure of the world at large, the miniature apparatus constitutes a multiscalar medium of translation and knowledge-production. At some moments, Chaucer's *Astrolabe* even gives an impression of the technical object as a sort of extended organ of cognition and sensation.

Chaucer's language is certainly provocative. He employs quasi-literary figures, and with vivifying traces of prosopopoeia, the functional device practically comes alive. A gathering notion of the weird hybridity of technical matter starts to take shape, along with the idea that astrolabic science, in practical ways, portends more-than-human intelligence. We end up with a sense of the astrolabe as an expressive interface that articulates across ordinary boundaries. Arianna Borrelli has already defined "astrolabe" to include verbal expression and nonverbal mental concepts, practical procedures, visual diagrams, and physical objects working together, clearing the ground for further analysis of the multimodal object.[10] Taking a wide-angle view, I insist that the astrolabe is transected by other lines (e.g., mineral substance, numerical quantities, external motions, and animal metaphors), obtaining its manifold utility by conducting the diverse vectors of nature and culture, body and mind, past and present. The transformative effects manifest themselves in varieties of simultaneous and mutually constitutive *translatio*.

Accordingly, I will start with Chaucer's theory and practice of textual transmission in the *Astrolabe*. But, because an astrolabe is a technical ensemble that involves more than verbal constructions, the geometrical and physical facture must be reckoned with as well to show how magnitudes and motions are translated in the device. Here the form is at once immaterial (hewing to a mathematical ideal) and material (rendered in the artifact). Finally, the mechanism relies on seemingly excessive and animating figures that communicate across yet other ontological divisions. Mercurial metaphors (e.g., matrix and spider) reveal an uncanny side to the scientific matter and extend the ensemble to encompass a kind of imagined multispecies modality. What follows draws a thread through disparate empirical and rhetorical elements of a heterogeneous assemblage, conjoining figures from overlapping domains that co-constitute instrumentality.

Translation and Translingual Ecology

Chaucer relates his work to the regional and sociopolitical milieu of his day, deferring to "the king, that is lord of this langage" (45–46). And yet the scientific matter is hardly parochial, exhibiting instead a worldly orientation in respect of its acknowledged sources, terminological dependence, and portable knowledge. Drawing together international knowledge practices, Chaucer's technical apparatus opens up transhemispheric horizons in more than one way. The treatise fundamentally depended for its existence on transcontinental text networks and scientific cultures from around the medieval Mediterranean. At Chaucer's time, Latin was still the default medium of technical knowledge, not yet English, and his translation efforts belong to the history of the slow vernacularization of textbook science. Aware of the contingencies involved, Chaucer speaks only of the limited adequacy of his English to accompany "a suffisant astrolabie as for oure orizonte compowned after the latitude of Oxenforde" (7–9). In his prologue, Chaucer adopts the pose of belated compiler: "I am but a lewde compilator of the labour of olde astrologiens and have it translatid in myn Englisshe oonly for thy doctrine" (49–50). A sense of just how much the work of translation absorbed Chaucer in manifold disciplines, traditions, and tongues—what Stephen McCluskey calls medieval astronomies and scientific cultures in the plural—can only be suggested here.[11] Specifically, Chaucer was involved in adapting a Latin translation of a work attributed to the eighth-century Jewish-Arab astronomer Māshā'allāh ibn Atharī, *De compositione et operacione astrolabii*. Chaucer also relays information from the *Tractatus de sphaera* by the thirteenth-century Parisian scholar John Sacrobosco, whose treatise, a staple of the liberal arts curriculum, passed on a version of antique cosmology and spherical geometry.

How was such a synthesis possible in the fourteenth century? The story begins with the abstract geometrical sophistication of Ptolemy's

Almagest, digested and developed into sophisticated planetary theories by scholars working around Baghdad, the Maghreb, and al-Andalus from the eighth century onward. Important conduits to the Latin West included Gerbert of Aurillac in the late tenth century (who returned from northern Spain with the abacus and Arabic numerals) and, in the twelfth century, Adelard of Bath (who studied with Arab scholars in the East and was responsible for translating Euclid's *Elements* and the tables of al-Kwarizmi, and composing an introduction to Abu Ma'Shar and a treatise *On the Use of the Astrolabe*). The available texts and techniques engaged a circle of scholars working in the West Midlands, where a keen interest in so-called "Saracen calculation" took hold.[12] Growing familiarity with devices (e.g., astrolabe, armillary sphere, and planetary equatorium) provoked the further spread of mathematical and empirical methods that had already informed the Arab world. It has been noted before that, in the translation movement sketched here, the intensity and reciprocity of cross-cultural interchange stands out against the background of crusading; it is as though instrument-learning created a space for convivial interchange.[13] An "astronomical corpus" took shape around itinerant bodies of knowledge whose horizons were spacious, whose historical and transcultural dimensions were nonexclusive. One example is British Library, MS Harley 3647, a fourteenth-century compilation from Paris, where Sacrobosco taught, containing Sacrobosco's *De sphaera,* Māshā'allāh's *Astrolabium,* Gerard of Cremona's *Theorica planetarum,* the Toledan *Tabulae,* and four astronomical treatises of Thābit ibn Qurra. Academic science was soon Englished in such compilations as Peterhouse, Cambridge, MS 75.1, which includes the unique copy of *The Equatorie of the Planetis* (a text that begins with a translation of the Arabic *bismillah,* "In the name of god pitos & merciable"), and in Trinity College, Cambridge, MS O.5.26, whose matter includes Ptolemy, Alkabucius, Māshā'allāh, and *The Newe Theorik of Planetis.*[14] In sum, Chaucer's *Astrolabe* represents a westering spread of multilingual knowledge, where historical currents and countercurrents came to flow, blend, and pool in the vernaculars.

That scientific culture was on the move is everywhere evident in Chaucer's *Astrolabe.* Emerging as a manifestly transcultural and cosmopolitan endeavor, the eventual confluence of astronomical practices that join the figurative and predictive, the descriptive and demonstrative, and the cosmological and computational modes is constitutive of the science. What Chaucer had ahead of him to translate was consequently predicated on linguistic migration and cross-cultural exchange. Just so, his theory of translation appears rather all-encompassing: "Suffise to the these trewe conclusions in Englissh as wel as sufficith to these noble clerkes Grekes these same conclusions in Grek; and to Arabiens in Arabik, and to Jewes in Ebrew, and to Latyn folk in Latyn; which Latyn folk had hem first out of othere diverse lan-gages, and written hem in her owne tunge, that is to seyn, in Latyn" (23–29). He says English is as adequate as Greek, Hebrew, or Latin for translating "trewe conclusions," a view that is remarkable for refusing to exalt one or another tongue. (Briefly compare two contemporaries. Thomas Usk, rather disingenuously given his own efforts to render parts of natural philosophy in the common tongue, takes a position that is the polar opposite to that of Chaucer: "Let than clerkes endyten in Latyn, for they have the propertie of science and the knowynge in that facultie; and lette Frenchmen in their Frenche also endyten their queynt termes, for it is kyndely to their mouthes; and let us shewe our fantasyes in suche wordes as we lerneden of our dames tonge." Giles of Rome remarks that philosophers invented Latin to better communi-cate "the natures of things, the customs and men, and the course of the stars," discouraging anything like vernacular astronomy.[15]) Whereas others would reassert linguistic hierarchies, Chaucer levels and histori-cizes them as local and effective varietals. Perhaps he went further in seeing that any vernacular is itself crisscrossed by others.

He had only to reflect on the fact that his English treatise is per-vaded by a rich technical vocabulary descending from regions between the Persian Gulf and the North Atlantic to know that the sit-uation was complex. A barbarolexis of borrowed terms in astronomy (Arabic *azimuth, almucantar, alidade, zenith,* and *nadir*; Greek *horizon, planet,* and *zodiac*; Latin *mater, latitude,* and *umbra versa*) is expressive of the profound intimacy of neighboring scientific cultures—Judaic, Christian, and Islamic. Vernacularizing science texts often relied heav-ily on such lexical assimilation.[16] Numeration told a similar story. Chaucer is cognizant of his use of "noumbres of augrym" (124, 126), adopting numerical forms whose derivation is Hindu-Arabic.[17]

Indicating the still unusual nature of the system, English calendars sometimes included tables for converting Roman numerals to Arabic, and arithmetical treatises began with elementary numeration (describing new figures, including the cipher, and the place-value system). Some Latin and English translators who chose to set out numbers left to right also occasionally imitated the right-to-left directionality of Arabic mathematical notation. Chaucer also affectionately, if overenthusiastically, attributes names of several months to "statutes of lordes Arabiens, somme by othre lordes of Rome" (130–31), since fewer of them are actually derived from the former than he indicates. Chaucer defers to the Muslim forerunner "Alkabucius" (126) as much as to the Greek "Ptolome" (228), and hews throughout to the lessons of the Jewish-Arab Māshā'allāh, his chief source. In translating science, then, Chaucer produces a "vernacular cosmology" alert to the transcultural currents and energetic exchanges of language and learning on which astronomy had long depended.[18] In its nomenclature alone, the ecolinguistic texture of the treatise would have been obvious. It is not just that Chaucer's treatise avoids xenoglossia. Nor does he achieve mere diglossia, according to which different languages are socially stratified and allocated separate functions.[19] The *Astrolabe* materializes the sort of translingual ecology that has been associated with some of Chaucer's other important works.[20]

The ecumenical scope of the translated matter accords with contemporary developments, showing how Chaucer participated in layered text and translation networks. It is a literary inheritance that stands out against the monoculturalism of scientific publication today, which is the result of the eventual passage from plural and polyglot disciplines to ones that are now narrowly Anglophone.[21] It wasn't always so. Chaucer's instrument discourse models a cosmopolitan *translatio studii,* recollecting antecedent transcultural exchanges. Part of what his treatise imagines into being is the intermingling of cultures without antagonism, or at least absent a reflexive protectionism that can be detected in so many other places in the premodern past. It participates in linguistic and literary systems that, however learned (i.e., artificial and acquired through painstaking study), construct a sense of cosmopolitan possibility.[22] In this respect, perhaps we can say that the *Astrolabe* takes its place alongside other efforts to "deprovincialize" medieval knowledge practices.[23] At the same time, the device recalls to users that skilled knowledge is locally situated in practice, made for and within a particular time and place of understanding. The transmission of astronomical knowledge in language is only the most obvious example of an expansive sense of instrumentality. Verbal *translatio* has a corollary in other transformations specific to the science (mathematical and material), suggesting that, in both descriptive and functional modes the instrument is capable of expanding horizons.

COGNITION AND MULTISCALAR MODELS

Using an astrolabe involves executing a series of actions and calculations described in Chaucer's treatise, and yet those technical operations are not themselves only textual. They are several things besides. We can go on to specify nonverbal elements of the apparatus that enable other aspects of translation, bringing the faraway nearby. An effective physical interface, enabling the conjunction of textual and extratextual, is necessary for modeling the cosmos in any given instrument.[24] Manual devices such as the astrolabe, equatorium, navicula, globe, albion, and rectangulus lend material form and function to the phenomenal world so that information can be presented to sensation and cognition. Chaucer's *Astrolabe* shares age-old assumptions about the cognitive convenience of such intermediating devices, sophisticated design objects that work by embodying mathematical models. An astrolabe itself instrumentalizes an artificial geometrical pattern, forming a usable facsimile of the heavens. Ambient terrestrial and celestial phenomena (horizon, meridian, zenith, latitudes, star positions, and so on) can then be reckoned and rendered legible and negotiable. Motions and locations can be compiled and arranged expediently, picturing them effectively. An amateur will not immediately spot the outline of the firmament in the physical instrument, but in fact a proportional diagram of the sky above the horizon is inscribed on the movable device. It takes the effort of exercising what Borrelli aptly calls the "geometrical imagination" while handling the device to see how coordinates are projected onto the plane in a way that translates between technical object and environment.[25]

To be specific, astrophysical phenomena are made to inhere in an astrolabe's physical shape by means of stereographic projection. A projection model obtains points, lines, and angles from curvilinear space, tracing figures in scaled-down quantities against a flat ground. An astrolabe, in other words, takes selected features of the hemispheric vault and translates them onto a small surface, graphically distributing coordinates on the metal plate. The transformation of three-dimensional reality into figures is more sophisticated than the elementary figuration of plane geometry, where, without losing properties, angular lines become visibly congruent or symmetrical, conforming to Euclid's famous fourth postulate (i.e., all right angles are equal).[26] Instead, stereographic projection results in substantial area distortion. Circles are not all symmetrical; spaces are not equal. As Chaucer says of almucantars, "somme of hem semen parfit cercles and somme inparfit" (269–70). Nonetheless, invariant mathematical properties are retained in the proportions of lines and shapes on the face.

Another challenge is posed by the fact that the astrolabe intercalates three coordinate systems in the projection.[27] The first system is oriented around the local latitude of an observer (Oxford, in the case of Chaucer's *Astrolabe*), so that the metal plate shows a projection of the sky overhead, subdivided by intersecting lines spreading at constant intervals from horizon to zenith. A second, independent system inscribed on the plate is centered on the celestial North Pole. Nested circles trace the two tropics and the equator, lines of the compass run from side to side and top to bottom, and a hole in the center marks the pole. A third coordinate system is indicated by the movable rete sitting atop the plate, showing the visible stars and ecliptic circle (which indicates the course of the sun and the zodiac band), forming a fretwork rotating on the polar axis. When everything is assembled in a handy device complete with flat figures, movable parts, and scales on the border, the result is a compact but apparently lopsided interface. An astrolabe appears somewhat askew because *some* lines and circles are centered on local latitude (zenith) and others on the pole, while the ecliptic is eccentric to both and revolves. Distortion, however, is the effect of an accurate abstract geometry. The total effect of mapping the celestial vault is a dense meshwork for which, as we will see, Chaucer has recourse to interesting metaphorical figures. In any case, the astrolabe presents a functional interface that formalizes, rescales, and overlays information from various sources. To observers whose universe appears to rotate around the earth, the sun and moon progress in succession along similar paths, other planets travel at different speeds along individual routes, and the constellations of the zodiac rotate at a different pace. The astrolabe is a mimetic and manipulable medium that superposes the various coordinates on the face of the mechanism. The device configures a perception of formal regularity, rationalizing and miniaturizing the *machina mundi,* a Lucretian phrase that Sacrobosco—who is one of Chaucer's sources—was known to adapt to refer to the instrument, manifesting a cosmic machinery.[28]

Practical astronomy consequently yields a sense of *translatio* to complement the others we have considered: referring now neither to textual transmission nor to an attendant transferral of cosmopolitan culture, but rather to the transposition of spatiotemporal dimensions embodied in a physical design. An astrolabe is an inscriptive and calculative medium that enables information extraction, conversion, and display because of the way, materially speaking, it figures remote phenomena. Importantly, it communicates among them as well. Above all, the aim of the astrolabe is to obtain reliable and replicable results across great distances, registering correspondence where none was obvious, furnishing further and more literal evidence of the deprovincializing potential of the scientific practice. Applying a universal geometry, holding basic values constant across space and time, the astrolabe amounts to a technological and transcultural idealization of a common world. In a sense, the mechanism is the message.

Chaucer is alive to the implications, which he could hardly avoid in having devoted himself to such a well-traveled scientific apparatus. Such instruments are oriented outward, encouraging a peripatetic intelligence and instantiating a shared topology through a kind of concrete abstraction. Chaucer even goes through a series of exercises on hypothetical projections in foreign lands, using geometrical devices as vehicles. Conclusions 23–25 provide instructions in how to "prove evidently the latitude of any place in a regioun" (239). Conclusion 26 describes how an equatorial astrolabe would have to possess almucantars "streight as a lyne" (943–44). Moreover, the calendars he intended for the third part were devoted to value conversion and would assume the commensurability of number across cultures.[29] Number, interval, and ratio constitute the rare phenomena of providing shared grounds for otherwise diverse geographies and polities, as if to suggest that cultural forms at their most numerocentric could be the less ethnocentric. An astrolabe seems well designed to traverse boundaries, even if not many made it out of the study or a specific locale. They interface among far-flung places by design. These often very small portable devices (58: "so small an instrument portatif") could contain nested plates for several regions: the so-called Chaucer Astrolabe at the British Museum covers nearly 20 degrees latitude and includes plates for Oxford, Paris, Montpellier, Rome, Jerusalem, and Babylon. Other extant examples range from 23 to 48 degrees, from Lower Egypt to northern France.[30] An astrolabe realizes in form and function the virtual adjacency of localities and polities, horizontalizing and miniaturizing vast territories in its compass; it congeals within its promising, portable body the very idea of transferability.[31] Here it is as if astrolabic projection, mathematizing and mapping the cosmos, anticipates a Latourian flattened ontology. An astrolabe is not unlike a one-dimensional trail map whose coordinates *"maintain intact* a certain number of geometrical liaisons, appropriately called *constants"*; despite "total dissimilarity" between map and territory, one corresponds to the other in pragmatic ways.[32]

Technoscientific Hybrids

Chaucer's *Astrolabe* attests to the manner in which practical astronomy obtains information by deploying transmedial designs based in language, mathematics, schematic models, and physical facture. Among these, we can find trailing metaphors, often surprising figurative devices that adumbrate something more of the nature of the technical object and its outward orientation. To speak of such figurations is to conjure a suggestive etymological connection: deriving from *fingere,* meaning to shape, form, arrange, touch, or handle, *figure* calls to mind the way the instrument translates mathematical propositions into a material medium. Abstract measures take a shape that is amenable to the touch, configured so as to be grasped, handled, and fiddled. The assumption informs a subject I will return to later in the book when discussing the origin of *fictus* (perfect passive participle of *fingere*), suggesting something "technically fictional" involved in empirical measures and model-making. It is not that the information produced by such figures is false or unreal; on the contrary, one can observe with medieval commentators that even positive proofs, whose value straddles the rhetorical and veridical, are always dependent on technical inventions. Observational methods require considerable artifice and innovation, and instrumental objects sometimes reveal themselves to have a very large imaginative scope.

An instructive example to compare with the astrolabe is the contemporary timekeeping device known as the navicula, a horary instrument that captures solar or stellar altitude in order to render the hours on a picturesque ship-shaped dial (Figure 2). Its main components are a central mast with sliding collar, from which hangs a string with a bead and plumb bob, as if forming part of the ship's loose rigging; two symmetrical castles with pinhole sights, which may be said to simulate lookouts; and a deep hull etched with relevant scales and signs. To tell the time during the day, one works with the small vessel this way: move the collar up the mast to match the present latitude; swivel the mast so that the keel lines up with the correct zodiac date at the bottom; slide the bead on the string and place it on the line for the current date on the right edge of the hull; tilt the whole device so that the rays of the sun pass through the pinholes; finally, the plumbline should fall on the correct hour or half-hour. The navicula's origins are unclear. That it takes the form of a sailing vessel seems rather freighted with possibilities, suggesting among other things the ability of the user to navigate a fluctuating phase-space as though crossing choppy seas. It is perhaps an intrinsic part of the figuration, or scientific fiction, that one could somehow float through time and space: one is expected to manipulate the parts *pro gubernacione mali* ("for gouernans of Þo maste" in the Middle English), which is to use a nautical term for navigation. One becomes a temporary *gubernator,* a pilot of the ship dial by setting an accurate course. Alluding to a vehicle of transport, a navicula is a mobile medium that may produce various sensations and notions about the mutability of time and space, and which could reckon with the business of late-medieval port cities. Design elements do not seem incidental; they are, just because ornamental, expedient and intensely communicative.[33]

The physical shape and perception of an astrolabe gives rise to yet stranger coordinating figures, anthropomorphic and zoomorphic,

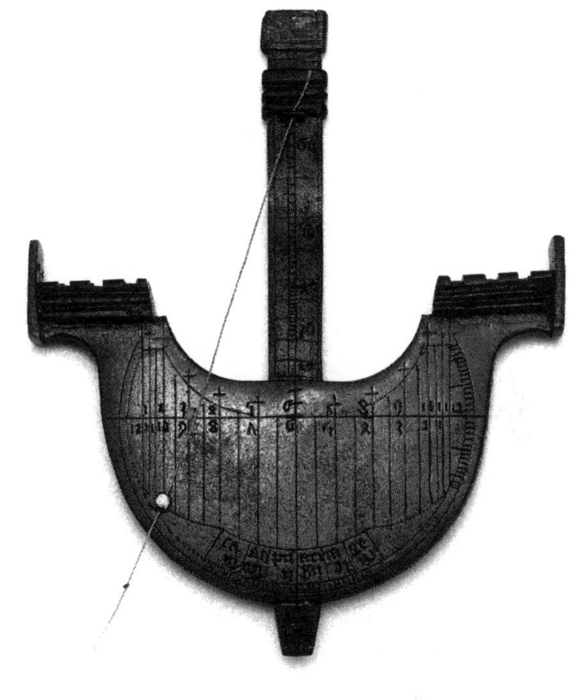

FIGURE 2. Navicula de Venetiis, circa. 1425–75 (National Maritime Museum, Greenwich, AST1146). Copyright National Maritime Museum, Greenwich, London.

whose chimerical presences are felt within Chaucer's *Astrolabe*. What he refers to as "figures and the membres of thyn astrolabie" (53–54) open up further dimensions to scientific understanding, and moreover situate both human and nonhuman species within a larger instrumental condition. Translating, they bespeak the hybrid nature of the technical object. Images that rear into view here include the wedge-like pin that goes through the middle and cinches parts together and is "clepid the hors" (197) and the zodiac band, or what Chaucer calls "the cercle of the bestes" (336).[34] Star pointers on the rete are sometimes "disposed in signes of bestes, or shape like bestes" (342), and surviving instruments in the Chaucerian style incorporate such figures as a hound's head, bird beaks, serpent tails, and human faces.[35]

A marvel to behold, the astrolabe inspired two especially dominant images over which we can pause. Translating for an amateur, Chaucer first of all speaks of the circular body of the instrument hollowed out on one side as the "moder" (Figure 3). The recessed interior is said to be "perced with a large hool that receiveth in hir wombe the thynne plates compowned for diverse clymate" (96–97). Various lines, numbers, letters, and names, including figures on the front (the "wombe side," 207), produce the appearance of "the werke of a wommans calle" (285), a term that refers to a hairnet but could also apply to the reticulated membrane of an egg or internal organ.[36] Chaucer did not invent these evocative terms; for instance, the English *moder* descended from Latin *mater* and Arabic *al umm*. It is as if we are hearing something more like the collective enunciation of the capacities of an ancient technical imagination. The concept is conveyed in and through multilingual corpora, and as such, the improbable anatomical metaphor is now more evidence of *translatio*. (It is a weird conceit with a long history indeed: everything from Plato's primordial *chora* to the *motherboard* of computers today suggests that the matrix exerts a strong grip on the technoscientific imagination.)[37] The allusive troping of technology at once evokes an organic reproductive space and an inorganic interface; it renders the material design of the instrument a vital matrix, enlivening the scientific object through *organopoiia*.

In the immediate context, Chaucer's *Astrolabe* is directed at the education of little Lewis, presenting the boy with a sophisticated learning apparatus, linking technology with pedagogy, so that it can be taken to tell a story about the radical dependency of living processes on surrogate matter—the mother. Adapted to the capabilities of a young amateur, Chaucer's treatise represents a variety of children's literature that trains up the intellect and engages affect, appealing to the curiosity of a child and expressing care for his development. The occasion is reminiscent of one earlier composition, a treatise on the astrolabe Adelard composed for a young Henry Plantagenet.[38] The idea that a child could be fostered on natural sciences is not a fanciful modern conceit. Four known manuscripts give Chaucer's *Astrolabe* the vivid title "Brede and Milke for Children," identifying the matter of the treatise with digestible foodstuffs.

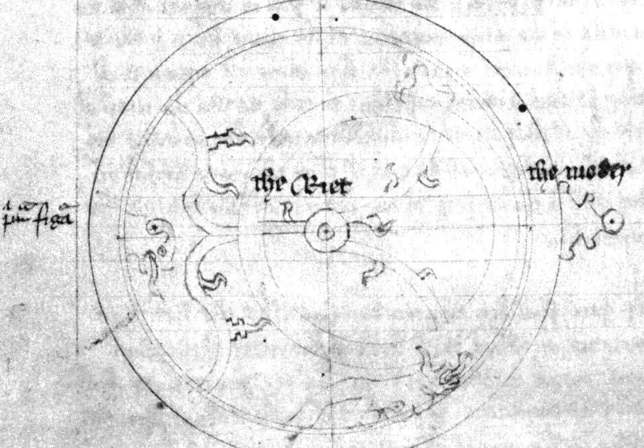

FIGURE 3. Front of an astrolabe as depicted in Bodleian Library, labeling the "riete" and "moder" and showing zoomorphic star pointers. Courtesy of the Bodleian Library.

Vital personal and impersonal agencies and attributes of the sciences are illuminated via what Jane Bennett calls a "touch of anthropomorphism."[39] It opens an epigenetic frame where we can start to see various sources at work in systems of knowledge production. Granted, what those humanizing gestures in the text imply for scientific inquiry has occasioned some debate among Chaucer scholars. We might be persuaded that the quasi-animate imagery facilitates a sense of belonging within a patrilineal intellectual heritage, situating the son in relation to a familiar masculine authority.[40] Yet the maternal language suggests something else at work. As Jenna Mead convincingly argues, "the resonance of the instrument's outer plate ('moder') maintains a powerful allure so that we should probably consider alternative constructions of authorship implied by the trope."[41] The terminology Chaucer is borrowing makes a deeper historical and technographic analysis available. Chaucer's astrolabe comes at the end of a long selection process that has taken place, quite without his involvement, in a fecund milieu (again, indicated by the Latin *mater* and Arabic *al umm*). His debts are plainly evident. Taking in much longer scales of development, the metal body of the instrument is proof of co-adaptive interactions of remote entities and environments. Made from mineral substance, coming from what medieval writers would also figure as a gravid mother earth, the astrolabe is embedded in geological time.[42] Alert to what a metaphor carries and delivers, such insistent figures may work to confound distinctions between animate and inanimate, between person and thing, positing a dynamic and diffuse morphogenesis. The metallic substrate is allusive of Manuel De Landa's "geological history" or Jussi Parikka's "mineral durations," terms that describe the planetary scales and substance of media technology to suggest that devices present matter that ultimately predates humans.[43] Tarrying with a matrixial metaphor is to resist a single, virile source. Here there can be no hylomorphic presumption of masculine mastery; there are rather plural, epigenetic points of origin. A scientific instrument is the result of so many intersecting lines that converge on a contingent point. It is from this perspective that the writer's plea of unoriginality should be heard as a strong disavowal of any notion of a lone progenitor: "I ne usurpe not to have foundenn this werke of my labour or of myn engyn" (48–49). The final word here can evoke technical and sexual propagation of a more ingenious kind: *engyn,* "skill" or "ingenuity," has origins in Latin *gignere,* "to beget," producing words that are close cousins, *genital* and *progeny.*[44] Accounting for his derivative work, Chaucer stands in a relation of surrogacy to the deep historical sources of the science, textuality, and technology. Considering the various geosocial matrices in which he is absorbed, Chaucer is rather like an episode in the life cycle of the apparatus, which is how technical objects engender modes of existence. Technical life is ever "transindividual."[45]

By now, the strain of *mother* and *womb* is surely felt, but there are other animating figures worth pausing over to get at what is excessive and exteriorizing about the astrolabe. Alert to the web-like formations of astrolabe plate and rete, Chaucer follows his sources again in speaking of parts of the instrument as evocatively spiderlike: the azimuth lines on the plate form "a maner croked strikes like to the clawes of a loppe [spider]" (276), and twice the rete is said to be "shapen in manere of a nett or of a lopwebbe after the olde descripicioun" (297, 299), recalling the traditional names of the components in Arabic (*ankabut*) and Latin (*aranea*). It is imagery that has proven as sticky as a cobweb (Figure 4). Ptolemy's *Planisphaerium,* a treatise from the second century CE on the flattening of the sphere that survives only in Arabic, refers to part of one horary instrument as a "spider."[46] Vitruvius mentions astronomical inventions he calls *arachne* and *conarachne,* which seem to refer to flat and conical sundials that are ancestors of the astrolabe.[47] As far back as the fourth century BCE, Eudoxus was supposed to have come up with a "geometrical spider" consisting of what appears to be a model of the heavens in the form of intersecting lines arrayed on a plane or bowl. The image may extend to astronomical tables Chaucer had meant to include: these were once called *zij,* deriving from Persian for "threads" or "chords," associated in early etymology with textile-making and coming later to describe intertwined rows and columns of data.[48] Spider and thread metaphors, crude though they may seem, are shorthand for complex material supports and tactics. They also introduce unexpected affiliations. Chaucer's "loppe" is no biblical spider—a negative exemplum of the wicked who entrap the innocent (Isaiah 59)—but instead conjures the figure of natural industry, practical intelligence, careful attentiveness, and good nurture set forth in the medieval bestiaries and encyclopedias.[49] The pregnant spider can also serve to reinforce the link between reproductivity and instrumentality. Albertus Magnus observes that spiders make webs into wombs, laying eggs in a kind of external uterus.[50] Bartholomaeus Anglicus says that one of the wonders of the spider is that she contains sufficient matter in her womb to spin a great web.[51] The spider is more often esteemed for its geometrical sophistication, an ability to join a web at regular intervals and right angles, distributing lines "yliche ferre fro þe myddel poynt."[52] The spider makes and maintains artificial structures and abstract figures, and even, as Pliny observed, deploys its body as a controlling plumb bob.[53] Aristotle, Philo of Alexandria, and Seneca all remarked the spider's knack for architectural designs; Ovid associates the spider with arts and crafts.[54] The spider, like the bee who fabricates hexagonal cells, possesses such formidable intelligence and instruments, unaided by human learning, that it could pose a challenge to human exceptionalism.[55] Moreover, the spider thwarts assumed gender hierarchies insofar as it was thought that the female, generating filament from within her body, is the preeminent worker and provider. Albertus Magnus observes that, as the evident doyenne, she has greater advantage and independence as compared to the male, whom she sexually dominates.[56] In sum, the feminized spider is a natural observer whose model the sciences belatedly and laboriously follow. She is like an apparitional image of the skillful and well-equipped technician. Hers is the matrix in which human instrumentation is raveled because she possesses the original *instrumenta*.[57]

Spider-like-ness has long been a summons to distribute sentience and skill across a multispecies spectrum, thereby contemplating a shared condition of instrumentality. In particular, the spider realizes a technoscientific ideal according to which one is not just observing an environment, but setting out threads through which external things are captured. Radiating lines, chords, angles, ligatures, and so on constitute a tissue of mimetic relations. Jakob von Uexküll's account of the spiderweb's correlation with the "image of the fly" is an instructive analogue: "The body structure of the spider has taken on certain of the fly's characteristics," which the web expresses and captures in a complex tracery.[58] The interwoven result is not mental so much as filamental. Still more suggestive is the account of pragmatic experience

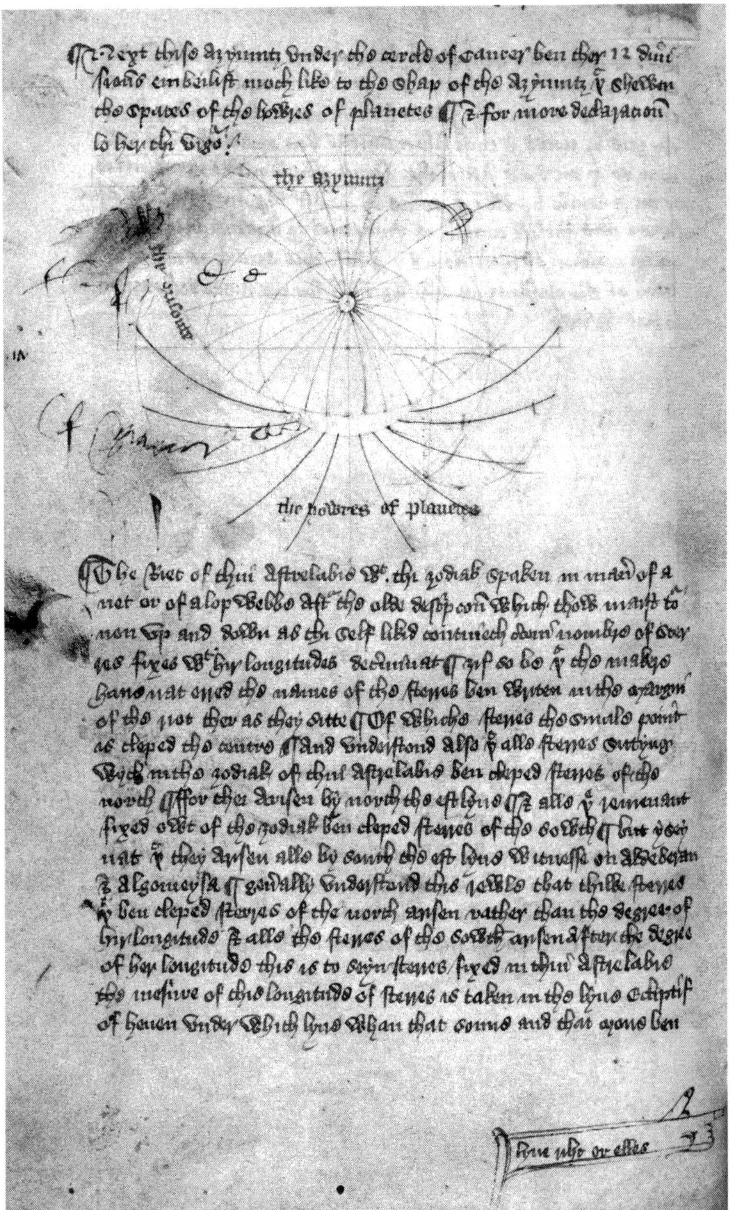

FIGURE 4. Diagram of an astrolabe plate showing spider-like azimuths (direction), zenith, pole, tropics, horizon, and hours. Bodleian Library, MS e. Mus. 54, fol. 8v. Courtesy of the Bodleian Library.

in Tim Ingold's SPIDER: "Skilled Practice Involves Developmentally Embodied Responsiveness." Practice relies on a flexible webwork, a set of relays enabling agile receptivity rather than the mere matter of a representation.[59] The way things are threaded through each other takes us beyond a static fly image to elucidate what might be dynamically transmitted by means of material affiliations. It is what ingenious figurative devices (whether cobwebs, mechanical retes, or metaphors) continually do by crossing species and substances; they are buggy metaphors connecting to essential features of a more-than-human technics.[60] The spider crops up repeatedly to suggest users are always caught, situated, and entangled.

Are we to think human practitioners become arachnoid in the astrolabe? Not exactly, but the technical fiction is nonetheless instructive for being so estranging. Here we can take it that the *instrumenta sensum*, to use venerable language for the receptivity of the organic body of the human (as mentioned in the introduction), are augmented by the artificial means of spider-like bodies. An astrolabe is, after all, an artificial prosthesis that intensifies and exteriorizes sentience and skill by arithmetical, alphabetical, visual, and haptic means, effectively producing variation on what Andy Clark calls the "mind-body scaffolding."[61] Bacon clarified for thirteenth-century readers that an astronomer's instruments have always amounted to a comparable sensorial extension and intensification of intellect. Speaking of the faculty of vision specifically, Bacon explained how heavenly objects are seen "per instrumenta visualia," their qualities and quantities confirmed "by the methods of viewing them with instruments."[62] In the textual and technical apparatus of the astrolabe, planetary phenomena achieve visible, legible, nucleated form in a virtual organ of sense and signification, an external technical object that distributes human corporal experience and cognition beyond the space of the so-called "natural" body.

A figurative womb or spiderweb is a minor conceit that may seem easy to dismiss when taking a practical approach to the contents of Chaucer's *Astrolabe,* but I hope to have demonstrated that, in the treatise, such meager fictions of science capture additional pragmatic information by metaphorical transformation, returning complex technical and mathematical matters to ordinary language. Empirical and ornamental properties of the science operate, in a profound sense, as one. The metaphorics of the device create unlikely associations and are useful insofar as they are implemented to measure the material and multidimensional cosmos in which practitioners are caught up. So many animal images are in fact projected within a beguiling matrix, employing intermediaries to make the world intelligible, and moreover recall us to a common instrumental condition. Seemingly accidental figures become models or, better, translational devices. They are instrumental to the meaning of astrolabic science.

An astrolabe yields a way of approaching technical objects and operations that can carry over to how we study a range of working devices, both medieval and modern. What other examples do we have of things becoming not merely applied but also, according to the old nomenclature, speculative instruments of philosophy (*philosophiae instrumenta*)? How might devices besides physical models portend the intensification and extension of sense and cognition in human knowledge practices? Which instruments inform and constitute the various arts? As we will see, many other phenomena open onto thinking about the prosthetic, imaginative, and even excessive modalities of instruments, whether considering physical configurations, textual and visual inscriptions, or learned techniques and intellectual systems that sustain the whole ensemble. A capacity to work is not limited to a physical interface; it is owing to what we can call the coordinated efforts of devices, discourses, and disciplines that anyone can be equipped to think or act. That is why we will next turn to graphic media inscriptions—that is, words and diagrams whose most reducible components are the point, line, and curve—in consideration of the proficiencies of figure-laden technical documents. We have already seen how acts of translation and textual transmission orient instrument users, presenting flattened figures to compose a view of the mobile universe. The argument proceeds to show that bookish matter can take instrumental forms too, mobilizing signs and orienting practitioners: aiding perception, modeling theories, shaping actions, and transmitting the reasoning mind to the furthest reaches. We will come closer to appreciating how language and literary artifacts belong within a larger, interdisciplinary apparatus, preparing the way for an eventual analysis of the liberal and mechanical arts. An immense variety of instructional and imaginative modes appear to converge in the concept of "instrumentality," that elastic term of art. The conclusion to which the evidence leads is that individual agency and intellect are contingent on so many assistive technologies, and that consequently we must expand what we mean by technical objects and operations. An ineliminable condition of intelligence is that we discover practical uses inherent in working devices, a recognition that can serve as a check against a reflexive modern anti-instrumentality.

GRAPHIC INTERFACES

Chaucer's *Astrolabe* shows how a manipulable model renders external phenomena intelligible, taking the measure of mobile terrestrial and celestial bodies. It sets out procedures for handling a tangible three-dimensional device, itself an embodied object of knowledge. But we saw that what is called an astrolabe is a multimodal technology whose technical properties are just as much inscribed as they are physically instantiated. Now we should press further to inquire into the affordances of surface inscriptions (i.e., point, line, and curve), and more broadly the proficiencies of graphic page design. How might two-dimensional verbal and visual media presented there serve as enabling epistemic devices, becoming not just representational or informational but also instrumental in ways we are coming to understand? To what extent do plane figures in books—alphabetic, numeric, or diagrammatic—function as technological interfaces, augmenting or advancing human knowledge? A comical scene in Chaucer's *House of Fame*[1] happens to suggest that such questions are neither recent nor just rhetorical, and that there has long been ambivalence about the value of inscriptive media with all its flat figures.

Consider the following medieval science fiction for what it suggests about the pragmatics of book-bound knowledge. In a wild flight of fancy, Chaucer appears in the poem to be seized by a golden eagle who imparts amateur lessons in physics and astronomy while carrying him close to the observed realities that would authorize those sciences. Blazing like "another sonne" (ln. 506), the bird portends the advent of radiant intelligence, conveying the mind to higher and higher realms, answering the poet's apparent desire for inspiration from the sun god, Apollo: "O God, of science and of lyghte" (ln. 1091). Yet the droll dream vision is full of ironic turns. Just as the eagle draws the poet up close to celestial phenomena and presents an unrivaled opportunity for direct access, the adventuring author demurs: "'Wilt thou lere of sterres aught?' / 'Nay certeynly,' quod I, 'right naught'" (lns. 993–94). The poet would rather rely on the written authorities:

> I leve as wel, so God me spede,
> Hem that write of this matere,
> As though I knew her places here;
> And eke they shynen here so bryghte,
> Hyt shulde shenden al my syghte,
> To loke on hem. (lns. 1012–17)

Chaucer would have readers believe he enjoys the comparative comforts of the study, situated at a remove from actual objects of scientific inquiry. It is a farcical moment, given how far he has traveled within the vicinity of the Milky Way, and would seem to close off avenues of inquiry in a willful withdrawal to the library. What should we make of his reluctance to look? It will strike many as a missed opportunity, impugning an unambitious or unscientific outlook. He appears to prefer the superficial to the solid, the flat to the spherical, and the graphically mediated to the real, leaving him ill-equipped to deal with the facts presented to him in such an obvious way.

Or is he? Inauspicious as it seems, Chaucer's satire runs in more than one direction and ends up illuminating something of the actual conditions of earthbound observers. For the dream vision is a featherbrained fantasy in which body and mind are ported through space and time toward the things themselves; that is the impossible dream. Chaucer's demurral consequently recalls the overambitious intellect back to the mundane necessity and effectiveness of human-scale inscriptions among other technical intermediaries. At stake are the epistemic means of acquiring and transmitting knowledge in a sublunary sphere in ways we can recognize as already instrumental, for example in the *Treatise on the Astrolabe*. As Chaucer will teach the young amateur Lewis, astronomy happens thanks to enabling graphic media and mechanical models that constitute the working apparatus. Antecedent works of Ptolemy, Māshā'allāh, and Sacrobosco are among the resources required for scientific practice, not the negation of it.

Technical texts are essential supports and would be said to reproduce in small what Chaucer elsewhere calls "thilke large book / Which that men clepe the hevene ywriten ... / With sterres" ("Man of Law's Tale," lns. 190–93). The point can be understood in a practical rather than esoteric sense. Among the necessary preconditions for useful knowledge are accessible texts, technologies, and technical know-how, which are any serious practitioner's pay for scientific understanding. In rejecting a lofty and unmediated vision of the stars, then, Chaucer counts himself among the well-equipped practitioners whose textual and technological intermediaries are indispensable to the production of knowledge.

Even Chaucer's dream vision is a kind of technopoetic apparatus, evidently tracing specific acts of an instrument operator in the text. Instrument use appears to inform his scientific imagination and collapse a distinction some are liable to make between, say, written work and fieldwork. In composing the *House of Fame,* effectively diagramming the heavens, Chaucer may have used a physical astrolabe to work out the celestial coordinates implied by the date of the dream, December 10.[2] He also knew how to check whether the constellations were arranged so that the sun rose over the horizon ahead of the "Eagle star Altair." Moreover, the fictionalized poet situates himself and the talkative eagle as if both were lodged in a particular area on the surface of an astrolabe. The eagle is akin to bird-shaped pointers on many astrolabes, part of the graphic design.[3] All this is to say that these literary fancies help us see how far textual, visual, and mechanical devices can become integrated into a full-featured, working technical ensemble.

It may help to recall that an astrolabe was once called a "book of brass" (*liber aeneus*), recalling the interrelation of media forms.[4] There are also examples of parchment and paper astrolabes with rotating discs sutured within instructional manuals.[5] Contemporary libraries kept records of books and instruments for borrowing (astrolabes, quadrants, spheres, and so on), suggesting something of their closeness.[6] In sum, the graphic elements found on physical plates and pages devoted to the astrophysical sciences were seen as having complementary technical properties and capacities. Both are part of the ensemble we are terming "instrumentality."

The reciprocal relations among textual signs, visual diagrams, and mechanisms are most recognizable in the pages of instructional manuals devoted to astrolabes, equatoria, naviculae, globes, and so on. That is where graphic inscription may be said to *underwrite* empirical observation and cognition. Many sorts of graphic artifacts could be explored to support the idea, but my attention here will be focused on how line-drawn geometrical diagrams take up and transform visual images into mental and manual operations, providing a fuller picture of situated knowledge practices that rely on plane figures and formats. It should become clear that inscribed points, lines, and curves—those constituent elements of graphic page design in many scientific documents—have a capacity to work as instruments. Presenting pragmatic inscriptions, they *document* knowledge in a strong sense that goes back to the Latin *documentum* (a model, proof, instruction, or written instrument), deriving from the verb *docere* (to teach, tell, or show).[7] In this formulation, the accent should fall on propagative acts of instructing, modeling, or evidencing in technical writing, stressing the interactive and iterative possibilities of flat media.

What follows first below surveys recent scholarship on medieval knowledge organization and visualization, examining the main affordances of diagrams and related figures. They include formats that were developed long ago but continue to play a leading part in intellectual culture. Then I turn to the pragmatics and thematics of something known as the "chord diagram," a foundational figure of mathematical geometry, geography, and astronomy. Here is scientific instrument whose durable functions can be tracked across time and space, not least because the chord remained such a useful device for orienting observers within time and space. The conclusion is that such superficies as diagrammatic figures are not inert marks, but rather transitive and transformative objects, graphic figures with technical facilities that enable movement between domains, made to equip the medieval scientific imagination.

Lines of Inquiry

The point, line, and curve are conspicuous elements of technical communication, but they need to be situated first of all within the "graphic field" that is the medieval manuscript page.[8] Handwritten documents are generally composed of rows of letters set on pricked and ruled guidelines, producing an image that is recognizable *as* text. Moreover, a good deal of the functionality of any document resides in the visual order and arrangement of the material, part of what Elaine Treharne terms "architextuality."[9] Recall that for Isidore of Seville, books are *instrumenta* because of the way information is presented. His *Etymologies* is disposed in such a way as to be especially serviceable: copies were given tables of contents, rubricated titles and subtitles, chapter and book divisions, and marginal annotations and corrections, and the whole is typically sequenced and sorted into twenty topics.[10] An obliging note to the table of contents indicates the care that would have gone into the arrangement: "So that you may quickly find what you are looking for in this work, this page reveals for you, reader, what matters the author of this volume discusses in the individual books."[11] Emily Steiner has explored how later medieval encyclopedic compilations experimented with paratextual finding aids (e.g., tables and alphabetical indexes) that have gone on to shape modern information technologies. They are innovative media forms that "activate a text, making it accessible, quotable, even compelling for readers."[12] Navigational aids remain effective and familiar elements of reference works.

Part of the value of any written document derives from its visual image and formatting, but texts can contain diagrammatic figures that are more distinctively instrumental and interactive: graphic elements of page design that may be movable, transferrable, iterable. Some diagrams represent useful philosophical *distinctiones* or a *divisio textus*.[13] Others within construction manuals are more hands-on: the user is drawn into acts of making, writing, numbering, graphing, and labeling a given surface. Making an equatorium, for example, requires a circular metal plate or board, preferably covered with parchment, upon which is inscribed the correct geometry and scales on the rim or limb: "this lymbe shaltow deuyde in 4 quarters by .2. diametral lynes in man*er* of the lymbe of a comune astrelabye & lok thy croys be trewe proued by geometrical *conclusion*."[14] Similarly, to make the navicula dial described in the last chapter, the reader is told to use rule and compass to compose geometrical lines: "First make a cercle and diuide hym by .A.C. and .B.D. diametres and the centre of tho myd point of tho cercle be called .O.," and so on.[15] Crafting and drafting techniques coordinate hand and eye in the motions of delineating a useful design, creating a work surface that is diagrammatic. Diagrams were often carefully drawn in manuals themselves, where they could serve as visual references or image transfers. They were graphical instruments. Such is most clearly the case with the geometrical drawings of the parts of a navicula in one manuscript where the sketched figures are labeled "instruments."[16] In the medieval period, two-dimensional figures traced on a flat page include a large and rich variety of practical infographics—charts, maps, lists, tables, calendars, marginalia, musical notation—that have various applications. Historians of art, science, and technology have taken a keen interest in those elements of technical communication, showing how knowledge is visualized and applied in different fields. The medieval evidence confirms that the point, line, and curve exhibit efficiencies in the aid of perception, cognition, and physical application.

At its most general, the diagrammatic function may be defined by the way a visual figure puts knowledge to a specific purpose, mobilizing information rather than merely representing it. A diagram is a graphic contrivance with practical consequences. What permits the figure to work at all is that it is a superficial projection of figures on a flat ground that can be taken up and used. An influential strand of scholarship has helped explain the expressive and functional aspects of such visualization systems.[17] As Valeria Giardino says in an analysis of the semiotics and pragmatics of figures, "diagrams . . . happen to be very good *inferential shortcuts* in problem solving. By externalizing the information in space, they reduce the amount of search required to find the relevant information and solve the task."[18] They are also not without their paradoxes. "Diagrams, written texts, graphs, and maps," observe Sybille Krämer and Christina Ljungberg, "are all graphic inscriptions on a formatted surface. This semiotic flattening is totally artificial. There are no two-dimensional bodies in our life world, and yet we treat surfaces with inscriptions and illustrations as if they had no deep structure, no behind or underneath."[19] And yet diagrams are no less productive for being so abstract and schematic, and in practice, their explanatory or operative power is based in strategic abstractions and deformations of observed phenomena.

Medievalists have lately provided valuable historical perspectives on the roles diagrams perform in various contexts where they constituted what Jeffrey F. Hamburger calls "tools for thinking."[20]

Concept-centered diagrams held an important place in the history of philosophy. Ayelet Even-Ezra has drawn attention to the pervasiveness of horizontal tree diagrams in scholastic culture, surveying their functions in the margins of Aristotle's *Organon*.[21] As we will see later, an evolving Aristotelian logical corpus formed part of a curriculum devoted to a relatively abstract instrument (*organon*) of thought. It is only fitting that most texts should be subtended by graphic instruments to orient scribes and readers alike as they moved through a dense course on logic.[22] Flat images were also important in astronomy, where they conjure external realia, though in these cases the diagrams may be more pictorial than the philosophical tree diagrams. Zodiac constellations have been treated as paradigm cases in selecting figures from a vast firmament to make pictographic signs that refer to changing states.[23] Another instructive case is that of an illustrated board game, which maps the heavens on the interactive space of gameplay. Steffen Bogen has analyzed examples from the thirteenth-century Castilian manuscript of King Alfonso, including both planetary chess and four-season chess. Both games involve a kind of recreational astronomy and induce players to set knowledge into motion on the board. In planetary chess, players move the planets around spheres and win or lose depending on the relations of zodiac signs, in which case gameplay "reenacts the way planets influence each other."[24] Chess requires the making of and playing with diagrams to be sure, leading Bogen to suggest that we "understand the *interaction* with any diagram as a kind of game."[25] Just as engaging with a chess board is discovering and executing possibilities within given play-space, so working with other kinds of diagram calls on cognition, inscription, and physical action. Medieval diagram culture developed many and diverse designs, suggesting a kind of whimsy or playfulness: forms were innovated and augmented all the time.[26]

Geographical maps are yet other figures that occasion subtle analyses of medieval diagram culture, illustrating how an artificial surface projection (e.g., Ptolemy's latitudes and longitudes) can produce and sustain useful knowledge. No map is its territory, and yet maps practically orient users within space and time.[27] Accordingly, Peter Barber and Catherine Delano-Smith discuss a set of abstract maps that are "diagrammatic in style": they are zonal maps, showing climates or other features of earth science, and *T-O* maps (*orbis terrarum,* the orb/ circle of the lands, with the emblem of a *T* inside an *O*) that segment earth into countries or populations.[28] The most well-known carto-graphic designs are large-scale *mappae mundi.* All such maps exercise the intellect and imagination to propel viewers outside of a situated space of viewing the graphic object, or at least orient them within abstract geographical configurations beyond the everyday. At least since Ptolemy's *Geography,* the world map was considered a marvel precisely for delineating the whole terrestrial sphere on a single pla-nar field. One medieval commentary remarks: "It is a great miracle how Ptolemy through his skill has brought under our eyes the entire round curve of the earthly world, as if he were depicting on a drawing table one single village."[29] Here points, lines, and curves capture the vastness, projecting a model of the whole in a compact display. In the interplay between map and mind, new perceptual and cognitive pos-sibilities can emerge, as Marcia Kupfer has observed: "A map closes the gap between exposing the structure of geographical reality in a single glance and imposing an organization on amorphous space, the visual template a prerequisite for further intellectual operations."[30] Outlining, projecting, rationalizing, and miniaturizing space and time for convenience, large-scale maps permit a viewer to traverse spatio-temporal dimensions otherwise beyond reach. In apparent contrast, local or regional maps would seem to restrict themselves to physically available quotidian spaces, but those maps too offer compact models that rely on the geometrical abstractions and simplifications. Matthew Boyd Goldie's analysis of local ground plans and charts of bounded geographical areas reveals the schematism involved in making small-scale maps workable and even walkable. A series of English ad-ministrative maps draw points, lines, and curves in a way that focuses attention on pragmatic systems and structures.[31]

In devotional contexts, diagrammatic figures can be no less imag-inative, interactive, and instrumental, as in the thirteenth-century itinerary maps from Matthew Paris's *Chronica maiora.* Separate col-umns show the route from London to Rome, depicting the individual towns and topographical features along the way, and for some regions included fold-outs with further information about places. The maps come without instructions, but Daniel K. Connolly most persuasively ties them to spiritual contemplation in the monastery, where the itinerary images could enable an imagined passage from place to place.[32] Hamburger has shown that much religious art is diagrammatic, developing logical systems and structures that are intricately patterned and serve as prompts to further meditation. Some of the visual formats were borrowed from logic, such as the "square of opposition" (contradictories, contraries, subcontraries, and subalterns), repurposed to teach salvation history.[33] Others draw on the visual language of trees, wheels, crosses, and so on, and very often they produce geometrical configurations of juxtaposing and interlocking elements, as in the typological patterns found in parchment and glass.[34] They are very likely to imply grids or other abstract arrays, as in the striking "metric relic" depicting Christ's side wound as a diamond shape against an implied grid. The diagrammatic wound is to be measured and meditated on for the protection of body and soul.[35] There are also the unusual inventions of Ramon Lull, the thirteenth-century scholar who took "the diagrammatic method to an unparalleled level of complexity."[36] Building on the formalism associated with logic and mathematics, Lull developed an expedient device to mechanize thought about divine existence, and in the process invented combinatorics.[37]

In many cases, diagrammatic thinking accelerates perceptual and cognitive operations, facilitating inferences in a way that natural language descriptions alone cannot, precisely because the graphic display enlists more-than-mental faculties. In short, it is a mode not only of thought but also of sensory and spatial experience. For example, a passage from the work of a fourteenth-century scholar and devotee of Lull justifies the use of visual figures on the basis of their apparent simultaneity to sight: "The reason I have had these visible figures, lies in that they are seen openly at once. Through seeing them, many things are called to the memory at the same time and at one time."[38] As Jill Larkin and Herbert Simon explain, a diagram formats data by visual "location" and supports "a large number of perceptual inferences, which are extremely easy for humans."[39] Giardino likewise observes: "By diagramming, humans recruit several systems that are already available for perception or action, such as the visuo-spatial system, the conceptual system, and the motor system, and establish an *external connection* between them."[40] The production and reception of diagrams calls on both mind and body in an interactive medium, as drawing and using figures is not just a visual but also often a heuristic and sensorimotor activity.

The practice can be tactile, as in the case of palpable codex-bound figures consisting of flaps and manipulable volvelles that facilitate "physical, multi-sensory as well as intellectual experience."[41] One of Lull's own figures takes the form of a layered volvelle that, no matter how cryptic his divine science, presents abstract combinatorics to the touch.[42] Paris's *Chronica maiora,* in which the aforementioned maps have expandable parchment flaps, also includes a computus volvelle for determining movable liturgical dates. "The practices of the *volvelle* incorporated the user's body as part of its operational mechanics," Connolly explains, requiring the user to spin disks or turn the whole book around to read information from the circular diagram on the page.[43] Diagnostic volvelles in medical books work in like manner but intensify the experience by explicitly referring to parts and operations of the body. The situation bears comparison with altimetric devices whose diagrammatic functions incorporate a functioning observer in diagramming a space of interaction. Technical diagrams showing practitioners how to measure distances and angles indicate that an observer's body is always implicated in the most basic observations. An operator's movements, visual mechanics, and physical dimensions must be taken into consideration; for instance, when using an astrolabe, one kind of measurement entails stationing the body at different intervals within a given area to take multiple readings. And measuring altitudes requires accounting for the observer's height, effectively plotting a diagram that includes *you* as a point in the geometricized area. As Goldie explains, "the eyes and the body become a part of the instruments of measurement."[44]

By now it should be clear that diagrams are inscribed spaces of intellection and action in which the viewer assimilates or manipulates figures, formulates or solves problems, acquires or applies information—facilitating skilled and situated knowledge practices that are embodied for cognitive convenience. The foregoing bears out Brian Rotman's idea of the diagram as "the work of the body," something that "call[s] attention to the materiality of all signs and of the corporeality of those who manipulate them."[45] Various handy and heady figures of mathematics count, particularly geometrical figures as construed within a vein of continental philosophy extending through Maurice Merleau-Ponty, Gilles Châtelet, and Gilles Deleuze. Addressing the kinematics of diagrams, Merleau-Ponty writes: "The subject of geometry is a motor subject."[46] Châtelet argues that diagrams, figures in virtual motion, "capture gestures mid-flight."[47] They supply, in short, the lines by which it is possible to move through space and time. Another suggestive example follows that will underscore the economy and effects of hand-drawn figures that were sometimes applied in the fields of geometry and spherical astronomy.

Mathematical Geometry, Plane and Spherical

Mathematical reasoning and representations rely on the point, line, and curve just as much if not more than do other disciplines. As Rotman argues, numbers themselves are "objects that result—that are capable of resulting—from an amalgam of two activities, thinking (imagining actions) and scribbling (making ideal marks), which are inseparable."[48] They are transitive instruments that construct, in figures, the conditions of mathematical thought.[49] Likewise, Danielle Macbeth observes that mathematical notation does not just display information, but also enables deliberating about numbers. As she says of the positional or place-value system of Arabic numeration developed in the medieval period, "collections of signs in mathematics do not merely *record* results; they actually *embody* the relevant bits of mathematical reasoning."[50] Mathematics is hardly static or wholly abstract; it rather relies on both inscriptive and intellectual transformations. A mathematician engages in "paper-and-pencil reasoning." Working with numerical values requires a workable sign-system, surface medium, and mental operations that are committed to the page, or rather on it.[51] We need look no further than the many visual proofs found in various quantitative disciplines, whether they are Porphyrean trees, Pythagorean tables, squares of opposition, rotational diagrams, or Euclidean geometrical propositions.[52]

Medieval mathematical geometry furnishes many examples of diagrams for working out concepts, visualizing problems, or demonstrating proofs. A beginning geometer may follow a sequence of steps to show the properties of a circle or triangle, and thereby form a new appreciation for the pragmatics of the figure. For instance, Roger Bacon describes how the sensorimotor experience of diagramming verifies the mathematical theory of the equilateral triangles. Referring to the first proposition of Euclid's *Elements*, Bacon says that the proof will hardly be accepted until the triangle is inscribed within two overlapping circles just so:

> The mind of one who has the most convincing proof in regard to the equilateral triangle will never cleave to the conclusion without experience, nor will he heed it, but will disregard it until experience is offered him by the intersection of two circles, from either intersection of which two lines may be drawn to the extremities of the given line; but then the man accepts the conclusion without any question.[53]

Demonstrations such as these are typically illustrated in the wide margins of copies of Euclid, occupying positions of importance equal to that of the text. Thomas Aquinas likewise speaks about the virtues of *diagrammata,* geometrical visualizations whose principles are "made known by discovery in the actual drawing of the figure. For geometers discover the truth which they seek by dividing lines and surfaces. And division brings into existence things which exist potentially."[54] In these reckonings a diagram is again no mere mimetic object or descriptive account, but an intellectual instrument through which geometry is realized, and a vehicle through which mental figures are apprehended and experienced by a sensorimotor subject.

A more sophisticated mathematical case study over which we can linger to demonstrate the semiotics and pragmatics of the point, line, and curve is that of the *chord diagram,* a figure that once lay at the bedrock of practical geography and astronomy. Chords have a long and distinguished history and were indeed the stock in trade of ancient and medieval astronomers. Chord theory originated with Hipparchus of Nicaea (150 BCE), putative inventor of the planispheric astrolabe, whose axioms helped Ptolemy some three hundred years later compose his influential chord tables.[55] Ptolemy tabulated chord lengths in sexagesimal fractions so that no laborious calculation was necessary for those who would come after him, as evidenced in tables of arcs and chords commonly found in the manuscript tradition. For example,

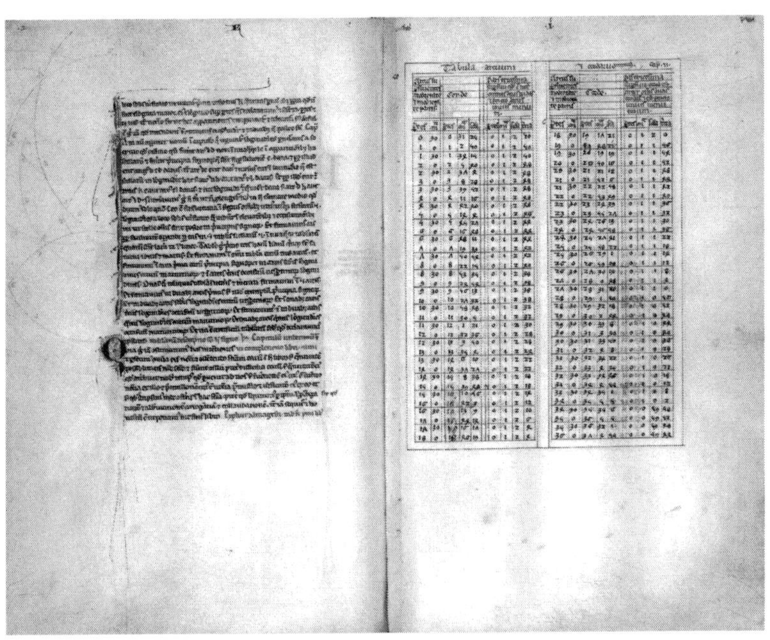

FIGURE 5. Chord tables from Gerard of Cremona's translation of Ptolemy's *Almagest* in San Marino, California, Huntington Library, MS HM 65, fol. 219r.

one could consult the chord tables from a copy of Gerard of Cremona's translation of Ptolemy's *Almagest* dated to 1279 showing lengths of chords of arcs in half-degree increments up to 180 degrees (Figure 5).

But behind the tables lay chord diagrams. How exactly is the geometry of the chord to be visualized? A chord is a line adjoining two points on a circle subtended by a central angle, whose computed values correspond to, and change with, the length of the line. A man can be seen drawing a diagram of the chord, emblematic of beginner geometry and astronomy, at the start of a section on measuring the earth, sun, and moon in a copy of *L'image du monde* (Figure 6). Today the properties of chords can be graphed and manipulated using digital tools (Figure 7). What is the function of a chord? A chord provides the means of plotting spherical coordinates to yield numerical values for figures, circles, and movements in space. It produces a flat figure that permits mathematical geometry to advance from the elementary *planimetria* (plane geometry) to the more complex *cosmimetria* (the

FIGURE 6. A man draws a chord diagram in *L'image du monde* in London, British Library, MS Harley 334, fol. 94v. Courtesy of British Library Board. Copyright The Granger Collection Ltd d/b/a GRANGER Historical Picture Archive.

measurements of three-dimensional mobile and spherical objects), to use Hugh of St. Victor's terms.[56] Chord diagrams can be used to measure points on a circle, or any arc of a sphere, such as may be done by calculating and coordinating the positions of celestial bodies orbiting earth on the Ptolemaic model. Here we must recall that, in medieval cosmology, the universe is treated as a series of nested spheres, with the planets orbiting around a central round earth. So situated, a practical astronomer will rely on chords to find such things as the distance of a star above the horizon or the distances between points in the sky, plotting them on the arc of the sphere along which they seem to be arranged. But, since it is not as easy to measure arcs as it is straight lines, the chord had been introduced to make sense of circumambient reality. A practitioner essentially identifies points and distances by solving triangles within curvilinear space, depending on tried-and-true trigonometric methods.

To get the gist of the chord diagram it is important to appreciate the instrumental value of the triangle, coming as it had to constitute a primordial figure animating the medieval sciences. Just as Euclid's equilateral triangle emerges from overlapping circles to demonstrate an elementary theorem, the Ptolemaic chord is a triangle inscribed within a single circle for a pragmatic purpose. It is notable that, above all, the triangle furnished a primary, transitive figure of thought and action in antique and medieval theory. As the mathematician and future pontiff Gerbert of Aurillac wrote on the subject, the equilateral

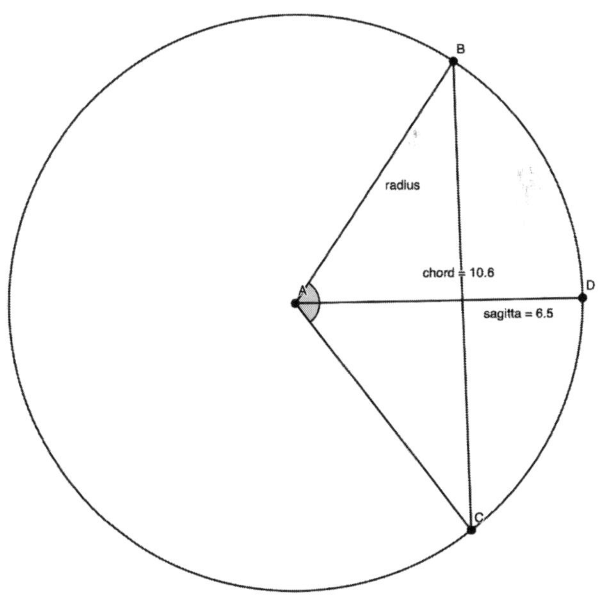

FIGURE 7. Line segments whose endpoints lie on a circle, subtended by a central angle, represent the chord whose ratios can be computed and tabulated. Created by author in GeoGebra.

triangle is "mother of all figures" ("matre omnium figurarum").[57] In his *Geometria,* Gerbert speaks of triangles as the grounding figure of plane geometry, which is the foundation of the rest of the science.[58] Writing a few generations later in his *Practical Geometry,* Hugh of St. Victor begins with the claim that the entire spherical universe can be delineated by means of right triangles for the expediency of human cognition:

> One must fit the shape of the world to this device. . . . [Those] vast spaces which our minds are not sufficient to comprehend, must be reduced, by rational procedures, to a small model, and so in a manner be made tractable to knowledge.[59]

The process of reduction appears to take on a triangular form in the mind, the large round world narrowed to a little point, rendered intelligible and subject to scientific scrutiny ("ad exiguum exemplar deducat, et sic quodammodo sub scientiam coarceat"). As the term trigonometry suggests, the figure of the triangle was for centuries fundamental to geometrical reasoning. And chords had to be mastered before anything else. They were the acknowledged means for managing the immensity of space, scaling external phenomena down to workable graphic models. The chord is therefore one of the most vital instruments of practical astronomy.

Accordingly, in Ptolemy's *Almagest,* the trigonometry of the chord is the first lesson of mathematical astronomy, from which subsequent theorems and practical applications flow. Several centuries later, Englishman Richard of Wallingford's *Quadripartitum* begins with the same elementary instruction. A fourteenth-century propaedeutic to spherical geometry and astronomy, *Quadripartitum* is a synthesis of mathematical principles drawn directly from Ptolemy's *Almagest.* Here the Greek language of the chord is combined with that of sines and cosines (*sinus rectus*; *sinus versus*; and *sinus duplatus*), the now more familiar terms developed by writers in Indian and Arab traditions to encompass the same teaching:

> The *sinus duplatus* is a straight line whose two ends coincide with the end of a part of the circle; and (the line and part of the circle) are

related as a chord to its arc. The *sinus rectus* is a half of the double chord with respect to half the arc. The *sinus versus* is always that part of the diameter which cuts both the *sinus duplatus* and its arc, and is like an arrow in relation to the chord and its arc.[60]

Geometrical figures, drawn by ruler and compass, illustrate the lesson in the opening leaves of the treatise, depicting arc, chord, and sagitta (Figure 8). Richard's lesson in chords circulated in manuscripts containing other works in the theory and practice of astronomy, and among treatises on the use of physical instruments. It was part of a longer education in how to work with visual diagrams and mechanical devices. One such ingenious device was the rectangulus (invented by Richard), which makes the two-dimensional diagram into a three-dimensional contraption to compute chords on the go. The apparatus incorporates rods, hinges, scales, and plumb bobs designed to point to objects and then read out the arcs of chords, essentially putting established theoretical principles into practice.[61] Here text and technical object follow one another, showing the mutually informing nature of the practical sciences in two and three dimensions, both on the page and in the physical mechanism. Such artificial models (imagined, inscribed, and mechanized) equip practitioners to comprehend otherwise intractable expanses. Point, line, and curve afford real knowledge; they put the mind in communication with faraway things.

MODEL DEPENDENCE AND METAPHORICAL TRANSFERENCE

The chord contains within it another and more profoundly figurative aspect—which is to say, a technically fictional quality that goes to inform contemporary instrumentality. Here the dimensions of flat figures expand in yet another direction, charting a path from the mathematical to the pictorial and eventually even poetical. We previously noted likenesses of the astrolabe (hairnet, webwork) and navicula (ship), emphasizing their reliance on such figments. Now, to speak of chords is also to imagine an abstract theorem in terms of an accessible and concretely embodied analogue, which is so often the way of medieval scientific models.[62] Why so? In an influential Aristotelian view, even the most abstract theory does not occur without simulated

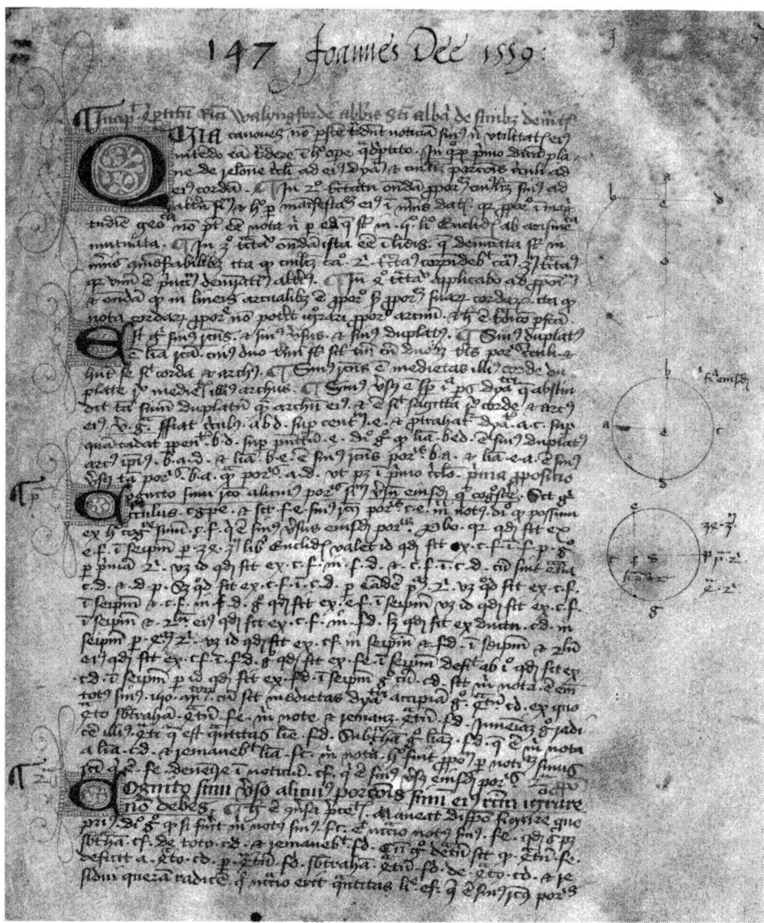

FIGURE 8. Prologue to Richard of Wallingford's *Quadripartitum* in Oxford, Bodleian Library, MS Digby 178, fol. 15r, showing at top right a faded diagram of arc, chord, and sagitta. Photograph by Bodleian Library.

images.[63] The assumption was axiomatic for later medieval practitioners: the sciences are substantiated by the use of expedient and vivid models: *formae; figmenta; exempla.* Aquinas held that, for angles, lines, points, and circles to be intelligible, they must be formed in the imagination.[64] In his treatise on the sphere, Nicole Oresme began with the idea that points and lines are not of nature ("in rerum natura"), but of the imagination ("sed solum fingitur per ymaginationem").[65] Elsewhere, Oresme speaks of the heuristic value of visualizing mathematical figures: "Although indivisible points, or lines, are non-existent, still it is necessary to feign them mathematically for the measures of things and for the understanding of their ratios."[66] The verb he uses in both cases is a conjugation of *fingo, fingere,* which can mean to form, to frame, to fashion, or to feign, and from which is derived the noun *fictio* for something fictional or fabricated. What has practical science to do with such artifice? Generalizing no doubt, John of Salisbury says all disciplines worked with *figmenta rationis,* conceptual fictions that go to inform rational concepts.[67] Exemplary figures or fictions, he explains, are instrumental to both learning and teaching. Scientific figures held in the mind, expressed in compass-and-line drawings, or embodied in technical objects reconstitute abstractions and aid comprehension.

Such is manifestly true of representations of the chord function, whether in visual diagrams or in accompanying tables or in explanatory treatments of them. Richard's description of the line as "like an arrow in relation to the chord and its arc" alludes to the way chords are a kind of imagined archery, something inherent in the mathematical idiom of *chorda, arcus,* and *sagitta.* Greek, Indian, Arabic, and Latin had long conceived of a line on the circle as string (*chorda*) to bow (*arcus*), and the perpendicular line segment an arrow (*sagitta*). Not only is the terminology of chords evidence of the spacious, transcultural horizons that we might be tempted to call "global," but it also constructs a particular relation to the space over the horizon by imagining it as virtual archery. The geometrical nomenclature and associated instruments preserve a hieroglyph of the common ranged weapon, by means of which a practitioner of spherical astronomy virtually shoots for the stars, equipped with fictive longbow and quiver.

Nor was the image incidental to geometrical reasoning. Bows and arrows had long been associated with elementary geometrical methods, the archer's tools and techniques called on to advance arguments and project diagrams. In a playful passage that involves a sagittarian solution, Gerbert of Aurillac employs what he calls a poetic figment ("poetarum figmento") to illustrate the Pythagorean theorem. If anyone wishes to determine a certain height, he explains punningly, here is clever means by which to discharge the problem ("modi jaculari ingenio investigare poteris"). Essentially, you shoot two bolts at different angles from one another:

> Get a bow, arrow, and line. One end of the line is attached to the tip of the arrow. Keep the other end in your hand. Shoot the arrow from the bow so that it hits the apex of the unknown height. Then tie the end of another line in the same way to an arrow or javelin. Throw either one, as you please, so that it hits the foot of the height, just as the first hit the apex. This done, pull back both lines...[68]

Lines attached to the two arrows form the base and the hypotenuse that can be used to calculate the unknown height ($a^2 + b^2 = c^2$), effectively diagramming open space. The procedure is neatly drawn in a copy of Gerbert's *Geometria* (British Library, MS Royal 15.B.ix, fol. 64r), where a small, elegant bow embedded within the text column sends into the margins two arrows wound with thread, as if the special attraction of the device were its capacity for visualizing otherwise imperceptible dimensions outside the text (Figure 9). The arrow was just one among other objects practical geometry recommended (e.g., sticks, mirrors, and shadows) to discover proportions or measure physical properties, configuring and concretizing mathematical ideas.[69]

The geometrical figure of the chord and, beyond that, the figurative imagery employed by other types of geometrical reasoning consequently present intriguing and sophisticated cases of model dependence, which, despite being indirect and abstract, nevertheless also rely on diagrammatic figures to work. Like other mathematical constructs, chords lie at the intersection of mental calculation, manual action, and metaphorical or conceptual figuration, blending technical proficiency with relevant imagery. Archery lends a degree of graphic specificity to the chord, calling on representational strategies to produce memorable and useful geometrical functions.[70] If mathematical

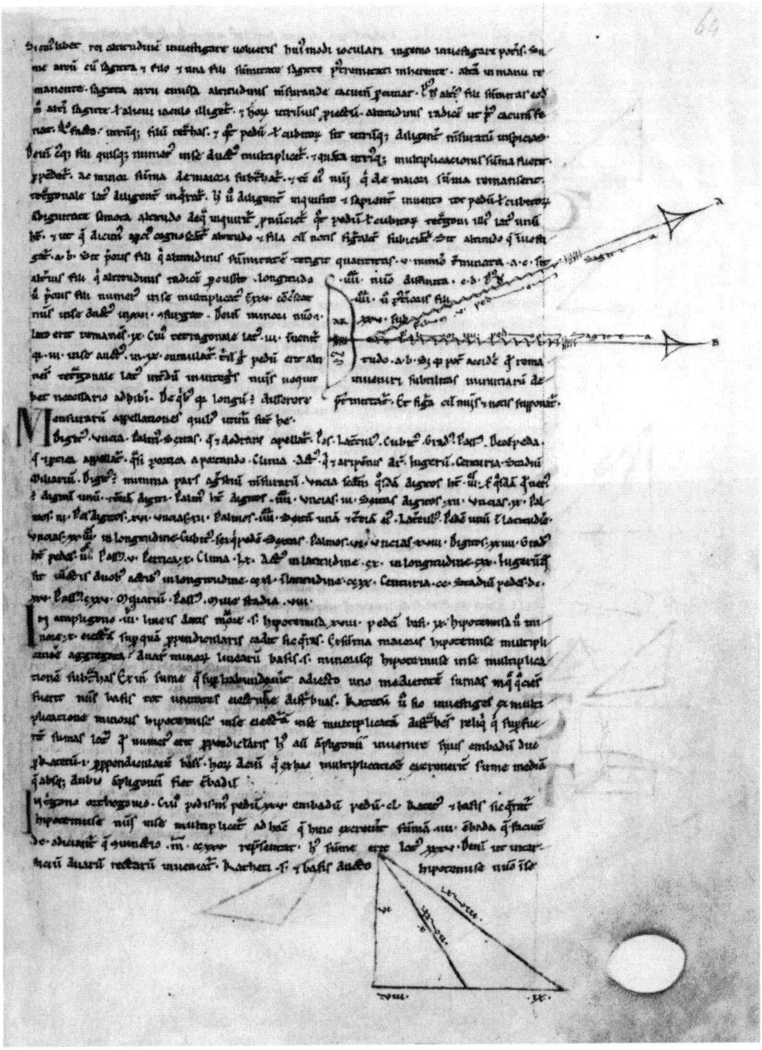

FIGURE 9. Demonstrating the Pythagorean theorem, a late twelfth-century copy of Gerbert's geometrical textbook depicts bow and arrows. From London, British Library, MS Royal 15.B.ix, fol. 64r. Copyright British Library Board.

inscription is always "a form of graphism,"[71] the chord diagram is just a more emphatic graphical instrument. It is a working model that must count as one of the most important and ubiquitous "image schemas" to have powered mathematical cognition.[72] Archery also recalls us to the artifice involved. Drawn on a flat surface, the diagrammed figure is paradigmatic in evoking a three-dimensional world that it aspires to comprehend and command, notwithstanding the humble two-dimensionality of the drawing. Recall that it is the nature of such figures to construct artificial, flattened figures that yield practical value just the same.[73] None of the selections, simplifications, abstractions, or distortions involved lessen the explanatory value of the diagram, and in fact they are constitutively practical. Just so, geometrical drawing and the reasoning flowing from it depends on the shared affordances of material artifacts and mental constructs commensurate to the empirical world.

Granted, a chord diagram is only ever an approximate, lossy medium and medieval observers were no doubt conscious of the limitations. Instrumentality is compatible with an epistemological modesty in recognizing when figures are, again, useful fictions, figures at a remove from the blinding things themselves (to recall Chaucer's *House of Fame*). Medieval theory was mindful of the limits of its models, which are avowed figments of *scientiae*. Chords are not exactly the arcs whose ratios they compute, as Richard for one candidly acknowledges at the end of his treatise on the rectangulus: "This method is quite sufficient for ordinary purposes, yet since—as Ptolemy proved in his first book of the *Almagest*—ratio of chord to chord is not exactly the same as that of arc to arc; . . . therefore there might happen to be some error in the minutes and seconds."[74] Chords, devices of convenience, facilitate ease of use by treating the curvilinear as rectilinear, triangulating points of a circle through a process of geometrical simplification or smoothing. They perform a kind of image transfer. Geometricization selects for regularities and resemblances, effectively reducing noise to boost the signal. Chords, like other models, amount to inscriptive and calculative devices adequate for accessing a cosmos addressed to human capacities. Yet, while they may be expedients, they are instruments capable of real work. Chords yield formal equalities that are not obvious to the naked eye, putting constraints on fluctuating phenomena, stabilizing fugitive phase states, figuring irregular circumambient movements, and in effect enabling earthbound intellects to rise to new heights.

FIGURES IN FLIGHT

The foregoing attests to the potency of the medieval scientific imagination instantiated in technical drawings and diagrams, replete with iterable objects that span the abstract and particular, the theoretical and practical, the mental and sensible. This is no less true of the bow-and-arrow figure, which draws the mind up and over, enabling improbable and inhuman flight across space and time, virtually transporting the practitioner; one moves over the horizon whose meaning, as Sacrobosco's *Tractatus de sphere* teaches, is *terminator visus,* "limiter of vision," something that mathematical calculation can artificially overcome.[75] Here we encounter a crucial feature of geo-astrophysical sciences revealed by the instrumentalized figure of the chord: the longbow, a ranged weapon whose shafts project over great distances, has always been exceptionally effective at touching what the hand cannot grasp and the eye cannot reach. As a familiar physical device, it is a relatively mundane tele-technology well suited to bringing the faraway nearby. Expressing the gravity-defying, extraterrestrial flight of the imagination, the bow and arrow can satisfy a desire to stand outside oneself without ever quite leaving the spot.

The sensation is one that would lead contemporary practitioners and dreamers alike to imagine themselves ascending to unlikely heights, scaling the known universe with the available instruments. Those working with flat maps, spherical globes, and rotating armillary spheres certainly indulged in an analogous and uncanny aerial fantasy, occupying an almost bird's-eye view of the whole. By constructing a world map, Ptolemy expressly aimed at grasping the entire terrestrial globe at a glance, and many who came after admired his achievements.[76] He was sometimes pictured holding aloft a celestial sphere as if affording a detached commanding view, a totalizing vision of the world (Figure 10). In a similar vein, in the allegory of geometry in the *Rothschild Canticles*—a richly illustrated florilegium of meditations and prayers circa 1300—a female figure representing the discipline uses a compass to measure a sphere as if to suggest that, by

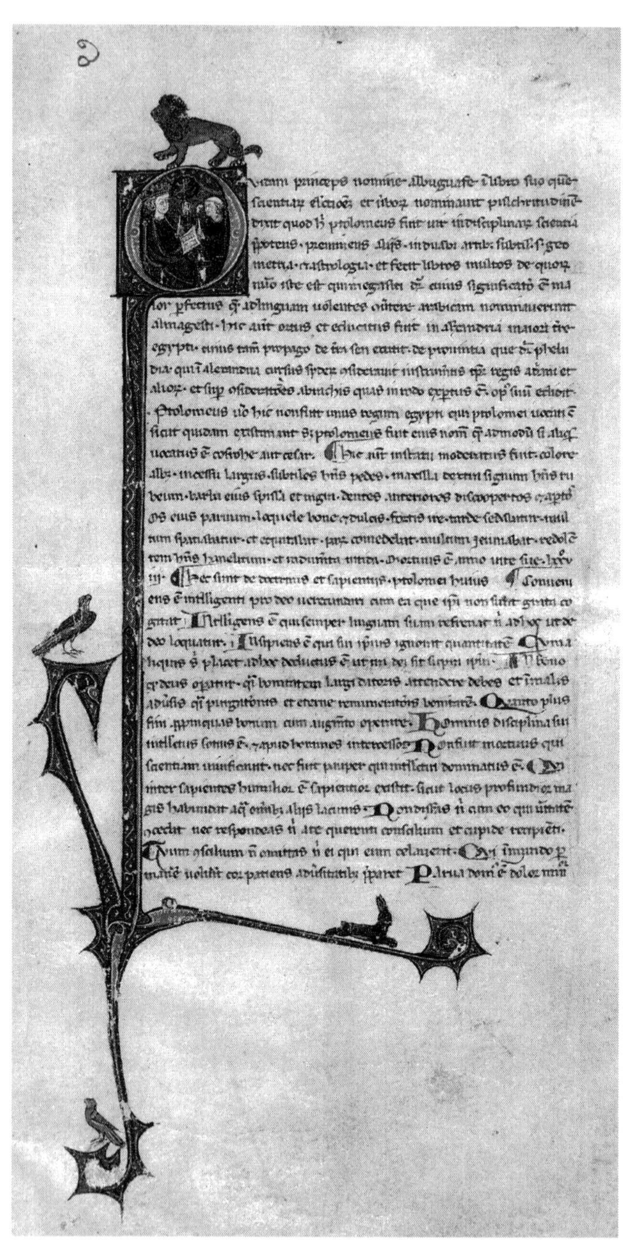

FIGURE 10. Ptolemy pointing to sphere in a historiated initial from Gerard of Cremona's preface to Ptolemy's *Almagest* in San Marino, California, Huntington Library, MS HM 65, fol. 1r.

her methods, she ascends to a synoptic perspective.[77] The same panoramic view was attributed to images of the Creator as Geometer, as seen, for example, in part of *L'image du monde* in British Library Harley 334, or in the famous frontispiece to the *Bible moralisée* in the Österreichische Nationalbibliothek.[78] Likewise, in the area of technical astronomy, points, lines, and curves of the chord enable mental images to project far outside the constraints of ordinary embodied human perception and motion; one achieves virtual overhead views. Flat figures constitute credible lines, even a nascent plotline, telling of extraordinary human transports and more-than-human transcendence. In imaginative literature, such a stance counts as an expression of an old rhetorical topos known as *kataskopos:* a familiar poetic and pictorial notion of hovering over a miniaturized cosmos, going back at least as far as the celebrated *Somnium Scipionis* and carrying forward in the aforementioned examples of the expert geometer and astronomer.[79] Medieval dream visions were consumed with such fantasies, soaring over human horizons, sometimes employing as vehicles birds rather than feathered arrows, a convention given innovative turns in Dante's *Commedia* and, as we saw, in Chaucer's *House of Fame*.[80] Chaucer is at his most sardonic and self-referential in the *House of Fame,* interrogating the merits of informational media that are never the *things themselves*. And yet he relies on technical intermediaries just the same in constructing an extravagantly strange but deft fictional apparatus that elevates the mind. Ultimately, one of the chief affordances of geometrical and astronomical instrumentation is the capacity to access otherwise impenetrable regions of understanding, just the sort in which contemporary poets would delight as a literary trope.

That artistic figures appear to pay tribute to technical objects and orientations should be enough to dispel the notion that there is some impassable gulf separating the instrumental from the intellectual and aesthetic. Key moments in the intellectual history we are tracing suggest instead that instrumentality is integrative and interdisciplinary, in fact a way of mobilizing knowledge across fields. And that is where we will head next: pursuing the constitutive interrelations among devices, documents, and disciplines within the *philosophiae instrumenta*. The medieval liberal arts curriculum exhibits a consistent concern with fundamental principles of utility in all manner of reasoned inquiry, employing methods that themselves rely on an array of mechanical tools and techniques furnishing the very conditions of intellectual life. Just so, we will talk about discrete technical objects and more about an instrumental orientation within the arts at large. Medieval discussions of such questions are explicit and exhibit a profound sense that instrumentality is an inherent property of human thought and action. What emerges from this analysis is that any wholesale critique of instrumentality—as pitted against supposed authentic or unalienated philosophical inclinations—will tend to defeat itself, neglecting to notice how profoundly dependent criticism itself is on a diverse and effective tool kit. We are in this book opening up a neglected area of inquiry for critical reflection and renewal, recovering a foundational instrumental attitude typically underrated in modern thought.

Learning Devices, or Instruments of Language and Literature

Where might we find clear formulations of the integration of the arts discipline under the rubric of medieval instrumentality? How far were technical objects and orientations seen as extending across epistemic domains, even informing the development of language and literary practice? And who might have explored the affinities between and among distinct knowledge practices, operative objects, and media technologies? In a notably eccentric author portrait, fourteenth-century London poet John Gower adopted bow-and-arrow imagery to memorialize his literary and intellectual ambitions, and the picture anticipates the directions such questions can take us given what has been said so far about the affordances of technical devices, diagrams, and documents. Gower's treatment of the issue limns a set of complex disciplinary norms that had preoccupied many thinkers: namely, the simultaneous integrity and interrelation of the arts and sciences within the acknowledged framework of the *philosophiae instrumenta*. His imagery is indeed unique for the way it visualizes a set of relations between poetry and philosophy, helping us to see how expansive the topic of this book could become for medieval theorists.

Three surviving copies of Gower's *Vox clamantis* contain large frontispiece images of a fashionable archer shooting at a suspended globe.[1] Each is headed by a short poem that begins "At the world I send my darts when I shoot arrows" ("Ad mundum mitto mea iacula, dumque sagitto"), evidently designed to represent a satirical posture (Figure 11). The text–image ensemble aligns with Gower's rhetorical and ethical imperatives, suggesting his words are like sharp projectiles, but further inferences can be drawn from the spatial dimensions and astrophysical abstractions. The frontispiece portrait has a recognizable geometric formalism, which is the hallmark of so much diagrammatic medieval art.[2] For one thing, the earth is rendered in the style of a *T-O* map, presenting an inducement to see in the whole picture an incipient diagram of the globe subject to the poet's ministrations. For another, in the disposition of *arcus, chorda,* and *sagitta,* the archer can be taken to outline a specific technical operation, alluding to the graphing of the chord function with which we are now familiar from the last chapter. Gower seems to conceive of himself as wielding the specialized instruments of science, in the role of a mathematical geometer no less, and as adventuring into neighboring domains of knowledge. Floating somewhere beyond the earth in the expanse of his (technical and poetical) imagination, a cosmonaut poet idealizes the overhead optics and mental extensions made possible by the instruments of geometry and astronomy, and he thereby engages the aforementioned literary dream of *kataskopos.* Those abstract flights of the imagination express recognizably Gowerian enthusiasms, a cross-disciplinary desire to command a total view of earthly experience. Gower's characteristic encyclopedism is an attempt to collect and catalogue, store and synthesize both theoretical and practical knowledge. One critic, for example, argues that the poet pursued a sophisticated synthesis of quantitative and qualitative learning in an integrative vision of the educated soul.[3] Another has dubbed Gower a "moral geometer."[4]

Apprehending the visual logic of the portrait, those who open the book will consequently detect not only a statement of the poet's learned authority (as the author of substantial works in Latin, French, and English), nor just an allusion to particular scientific devices (the chord function and divided globe), but also a long-standing preoccupation with the *divisio scientiae,* or branches of learning. Gower was fascinated with such higher-order concepts and cognitive technologies and had much to say about the way the arts disciplines are affiliated, divided, and deployed. For instance, in part of Gower's *Confessio amantis,* a hefty vernacular compilation spanning eight books of octosyllabic couplets, he lays out what he takes to be a comprehensive liberal education.[5] Summarizing the curriculum of the so-called School of Aristotle in which Alexander was tutored, the seventh book presents a detailed conspectus of its three branches of philosophy: *theorique, rhetorique,* and *practique.* The first comprehends the

FIGURE 11. Archer portrait in San Marino, California, Huntington Library, MS HM 150, fol. 13v.

mathematical sciences, including both geometry for measuring proportions ("Of lengthe, of brede, of depthe, of heythe"; 7.180) and the dimensions of celestial bodies ("The cercle and the circumference / Of everything unto the hevene"; 7.188–89), and astronomy for calculating magnitudes and motions in a dynamic three-dimensional field ("Which makth a man have knowlechinge / Of sterres in the firmament, / Figure, cercle, and movement"; 7.672–74). *Rhetorique* concerns rhetoric, grammar, and logic, while *practique* addresses ethics, economics, and policy. Intriguingly, Gower's lengthy metrical synopsis of the sciences in this book seems to be modeled on the lofty learning he ascribes to geometers and astronomers, who capture an overarching view in apprehending the mobility and sphericity of the cosmos. That is, Gower's poetry aims at something analogous to a comprehensive, nearly kataskopic vision of the world. Now returning to regard the memorial frontispiece image of the archer, the author portrait comes into focus as a transdisciplinary poetic figure embodying a holistic Aristotelian philosophy, showing himself to be well-equipped and engaged in practical reasoning. He becomes an emblem of the idealized integration of the arts and sciences, for which the chord and globe are metonymic of an implied *Organon*.

From a historical perspective, what is surely curious about the image of Gower adopting an instrumentalizing posture is the confluence within the same artistic project of what have become, in modern parlance, separate intellectual faculties with distinct epistemic norms. What has poetry to do with natural philosophy? Yet Gower's vision of overlapping knowledge practices is hardly novel, consolidating as he did the specialist subjects of the *trivium* (verbal arts: grammar, logic, and rhetoric) and *quadrivium* (mathematical arts: arithmetic, geometry, music, and astronomy) making up the *artes liberales,* in a tradition that had long branched off here and there to produce many variations on the scheme.[6] What follows in this chapter will recover lineaments of a theory in which the nature and order of the arts were explicitly debated with respect to their instrumentality. Arts disciplines were organized around core subjects, texts, and methods defined, most influentially in the twelfth century, by the Parisian scholar Hugh of St. Victor as *totius philosophiae instrumenta sunt.*[7] As we will see, the language runs deep and over centuries, establishing the basic parameters of educational theory communicated in what were known as didascalic treatises (or, introductions to the arts disciplines) circulating around the medieval Mediterranean. The concept was useful not least in supplying conditions for crossing or expanding disciplinary boundaries, including or excluding subjects, and allowing for the unhurried consideration of the significance of academic attainments. Early reflections were carried out largely in relation to Aristotle's *Organon* (*Instrumentum*), whether implemented in practice or codified in abstract curricular models like the one proposed by Gower. Elsewhere we noted the prevalence of diagrams as "tools of thought" in manuscript copies of Aristotle; we now turn to language and logic as cognitive instruments with manifold ends. The role of *logos* was indeed central, suggesting why medieval poets and many others might have taken special interest in the *philosophiae instrumenta*. Representing something of an epochal shift in thinking about the verbal arts, disciplines were increasingly understood to obtain unity and utility from lingual matter. Practically speaking, language and logic would seem to furnish the necessary technical and epistemic intermediaries of all subsequent disciplines. A major issue, then, was justifying the various language arts (e.g., grammar, rhetoric, and poetics) and an associated taxonomy of topics and canons of writings.[8] What I want to pursue first is the scaffolded nature of a general educational program across a range of examples, and the disputes they occasioned over the course to be taken through the arts, before coming around to consider the instrumental inventions of language and literature that fueled the imaginations of philosophers and poets like Gower.

Philosophiae Instrumenta

The custom of identifying and classifying the arts disciplines should not be mistaken for a uniform educational program across time, since it took various manifestations in different scholarly communities in the European Middle Ages. The liberal arts were born in controversy and remained works in progress. Scholars experimented with different taxonomies and texts, and the *trivium* and *quadrivium* were not always taken together or in the same order. In the eleventh and twelfth centuries, the main subjects were taught—often in piecemeal fashion—by celebrated masters attached to cathedral schools and monasteries. With the advent of universities organized into faculties, a more general educational ideal was structured around canonical texts and core subjects. Over their extended history, the liberal arts underwent innovations and were less a description of what was happening on the ground than a useful "heuristic tool" for analyzing the various branches of learning, instruments devoted to the description of other intellectual instruments, as we will see.[9] Nor did they always exist in harmony, as there were perennial quarrels over the roles and ranks of particular arts disciplines in different institutional settings.[10]

Nevertheless, an integrative and interdisciplinary ideal surfaced repeatedly in the literature despite, and sometimes on account of, scholastic strife and competing paradigms. A robust conceptual armature enabled a collegial vision to emerge in universities across Europe. Various inventories and introductions to the arts appeared that would describe in detail the requisite *philosophiae instrumenta,* and those descriptions included the writings of Hugh of St. Victor, Thierry of Chartres, Dominicus Gundissalinus, John of Salisbury, Robert Kilwardby, and John Gower, on down to Angelo Poliziano. One of the main points to observe is how cognitive instruments associated with mainline academic disciplines were the means both to construct and to critique, pressing on prevailing assumptions about the uses of higher learning. Instrumentality became a way of advocating for disciplinary identities, principles, and canons even as they were put into productive affiliation, discovering useful overlaps. It will become evident that writers undertook the hard work of mediating among internal norms *and* external relations, aiming at simultaneous coherence and epistemic convergence. There are recognizable tensions and pitfalls to any such cross-disciplinary reforms, then as now. The medieval university arose from a milieu in which, having to justify itself to itself (as we still must do), the arts disciplines were contested and (in today's idiom) ever in crisis. And yet an abiding sense of the value of learned instruments and technical expertise furnished a capacity for both internal critique and creative expansion.

Hugh of St. Victor's *Didascalicon* (ca. 1127) is exemplary in having articulated some of the principal assumptions and instrumental relations that would resonate among many other later medieval thinkers. In his sweeping panorama of the arts, disciplines are divided into theoretical, practical, mechanical, and logical. But, even as he schematizes in this way, he remains alert to the multiple ways in which one can describe the respective subjects, as indicated for example by his chosen construction *vel aliter*: "The theoretical is divided into theology, mathematics, and physics; or, put differently, into intellectible, intelligible, and natural knowledge; or still differently, into divine, instructional, and philological."[11] His is an effort to accommodate internal and external topics and disciplinary expertise, looking to construct a theory that knits them together. The most explicit means by which Hugh demonstrates the unity of the disciplines is in speaking of the arts as the tools of all philosophy, *instrumenta* conducing toward a common set of learning objectives.[12] He draws on the analogy of a journey to illustrate the process. Speaking of the *trivium* and *quadrivium*, Hugh plays on the root meaning of those terms to make a broader point about how each art should be pursued until its matter and methods can be taken elsewhere: tri-*vium* and quadri-*vium* are so-called because they are plural *viae*, or *paths to wisdom*.[13] In another place, Hugh clarifies that the arts strive toward the same destination, though not all run along the same course.[14] In treating the arts as instrumental, then, he is not saying that they are all the same. To use a modern colloquialism that has medieval precedent, for Hugh it is indeed often necessary to stay in your lane and avoid wandering (*errare*).[15] In an instructive passage, Hugh draws a distinction between teaching *of art* and *by means of art* ("agere de arte, et agere per artem").[16] For instance, grammar is a subject in itself, and yet other subjects can be treated grammatically. One error, then, is trying to demonstrate all the arts when teaching just one. And yet you would also go astray in failing to acknowledge related uses. In sum, the arts have internal coherence and yet should ultimately be approached, over a long course of study, as if constitutively interrelated. Finally, as his example suggests, grammar has obvious primacy in the way that all subsequent arts depend on an instrumental *logos*.[17] Here he develops an analogy whereby the disciplines are *parts* and *instruments* akin to limbs of the body, expanding the scope of the language arts in a manner that will parallel other developments in medieval thought.[18]

Language and literature indeed became central to the developing notion of instrumental disciplines, and we can point to specific factors that would make the connection more prominent over time. *Logos* was the main entry point for thinking about the interdependence of the liberal and mechanical arts, and the case only became stronger thanks to old and new thinking. Rhetoric, for example, had long been important to the major proponents, going back to the fifth century in Martianus Capella's *De nuptiis Philologiae et Mercurii,* which solemnized the so-called marriage of eloquence and wisdom. Philology personified is accompanied by her bridesmaids, who include the sisters Geometry (carrying the radius and globe and dressed in a diagrammed robe) and Astronomy (who emerges from a rotating orb of fire holding a measuring device and a book of calculations).[19] In this late antique allegory of higher learning, the intimate relations among the seven arts, their various technical objects on display, and the literariness of the work would solidify the sense that there is a deep correspondence between the disciplines under *logos.*

By the twelfth century, Martianus Capella's influence would be felt in Thierry of Chartres's *Heptateuch* (ca. 1140), where the marriage of Philology and Mercury is said to propagate the love of wisdom.[20] For Thierry, the love of wisdom is not some vague condition; it involves wielding the recognized core *instrumenta* of philosophy, which he calls "intellection" and "interpretation," and following a set curriculum through several subjects.[21] Though incomplete, Thierry's book was intended to be an immense omnibus of the instrumental arts: an anthology of textbooks on grammar, logic, and rhetoric (*trivium*) as well as mathematics (*quadrivium*). Lacking no scholarly ambition, Thierry names his *Heptateuch* the "instrument for philosophy as a whole,"[22] creating a one-stop shop for those who would pursue the *Organon.*

Thierry was the champion of an expanded *Organon.* He was already in receipt of ideas that had come down through the latest Latin translations of Aristotle out of Greek and Arabic, emphasizing the integrity and utility of grammar, rhetoric, and poetry among other arts. Here is another mainspring for the notion that language and literature belong to the instruments of philosophy. Previously for the Latins, just six core texts initiated a course in Aristotelian "logic," then understood as a demonstrative and dialectical art concerning the construction of valid statements. But Alexandrian and Arab commentators had enlarged the *Organon* to encompass seven and even eight works, adding

Aristotle's *Rhetoric* and *Poetics* to the list of required readings.[23] Now it would be understood that "logic" need not apply only to axioms and logical syllogisms. *Logos* ("reason" or "word" in Greek) had a linguistic expansiveness that allowed for the conception of rhetoric and poetry as equally fundamental and practical. How did those affiliated lingual instruments function, and what work were they capable of doing? Poetry, to use the newer phraseology, produces "acts of imagination." Rhetoric yields acts of persuasion.[24]

The notion that the language arts are essential and productive disciplines in their own right traveled identifiable routes in the later Middle Ages, coming to Latin-Christian writers via Arabic-Islamic sources. Abû Nasr al-Farabi (d. 951), an Islamic scholar working in the Middle East and central Asia, wrote pivotal treatises on rhetoric which instituted a philological orientation within neo-Aristotelian philosophy. His didascalic writings proposed that the verbal arts could be treated as instruments of academic study, government, and law, and contribute to public affairs.[25] Most influential in the European Middle Ages was al-Farabi's *Enumeration of the Sciences,* twice put into Latin in the twelfth century. In their respective translations of the work, both the well-traveled Italian Gerard of Cremona and the presumably Spanish Dominicus Gundissalinus (Domingo Gundisalvo) produced important distillations of Farabian philosophy. In initial chapters, for example, both follow al-Farabi closely in saying that knowledge begins with the *scientia lingue,* whose parts include not just diction and grammar but also versification. Then, rhetoric and poetics are listed among the eight parts of logic as per the Arab intellectual tradition.[26] It is precisely in this modernized, neo-Aristotelian *Organon* that we witness how evolving notions about the arts of logic and language were being digested in later educational theory. Rhetoric and poetry were elevated to none other than *instrumenta philosophiae.*

Gundissalinus's *De divisione philosophiae* (ca. 1150) is worth pausing over for the way Farabian ideas were seen to infiltrate and incline the arts more toward language and literature than ever before. We can detect some residual unease with the scheme even as he acknowledges the new settlement. Using terminology already familiar from Hugh of St. Victor, Gundissalinus explains what it means for an art to be both a "part" and an "instrument" of philosophy, framing a broader discussion of language and literary training.[27] As *part* of philosophy, each science has disciplinary norms and methods that are to be respected; as an *instrument*, each separate science can contribute to others. We recall the way grammar can be studied for its own sake or turned into method for others. The question Gundissalinus now entertains is whether the language arts, including rhetoric and poetics, are mere serving sciences, subordinate rather than constitutive philosophical parts. Initially he seems to treat rhetorical eloquence as a propaedeutic that lies outside of philosophy proper. All speech, after all, relies on nine "natural instruments"—the lips, teeth, tongue, and so on—to produce voice; and writing requires the "artificial instruments" of "hand, reed-pen, parchment, and ink."[28] Grammar simply employs such *instrumenta* to speak and write correctly; poetics has verse as its instrument, "a song composed in meter, created for pleasure or for a useful purpose"; and rhetoric has oration.[29] But he wants to know whether any of these verbal arts belong to higher learning, especially when it comes to pleasing meters and powerful orations. After expressing his doubts, Gundissalinus concludes that discourses both persuasive (rhetoric) and imaginative (poetics) are "species of syllogism."[30] They do constitute parts and instrument of the enhanced *Organon* passed down by the Arab Aristotelians.

As the twelfth century continued, others expressed similarly integrative visions that may have reflected the latest Greco-Arabic developments in Aristotelian philosophy. A well-known case is that of the English cleric and philosopher John of Salisbury. His *Metalogicon* (c. 1159), effectively an extended treatment of the *Organon,* takes logic in the most spacious sense to mean the study of speech. John lays out an ideal educational plan founded in words as *instrumenta.* In its most polemical passages, the work fends off detractors who say that the language arts are "useless." John counters that cognition itself depends on language training from a young age.[31] The arts augment a child's natural talent, or *ingenium,* characterized as the capacity to perceive, remember, and reason. These three faculties are given by nature as "both the foundations and the instruments of all the arts" ("omnium artium fundamenta et instrumenta natura"). So, one begins with a natural aptitude to study language and logic in the first place, thanks to various native instruments. In turn, language and logic become instrumental means to higher ends. Going on to form a more advanced philosophical understanding, the student will progress in the discovery that speech is the primary instrument of both rhetoric and dialectic.[32] Like others, John could see that each subdiscipline has its principles and methods on the path to higher learning. Accordingly, he distinguishes the means by which grammatical learning is incorporated: first by rhetoric, which uses words to persuade (e.g., oratory), and then by dialectic, which employs words to reason (e.g., the syllogism). One begins with verbal instruments in any case. In sum, "grammar prepares the mind to understand everything that can be taught in words," no matter the subject.[33] In this generous view of *logos*, each field of inquiry is separate while also standing in reciprocal relation to others. Moreover, the *trivium* is a sequential process culminating in a broad-minded civic education. What draws John to such a model? The interdependency and instrumentality of the arts serve as a check on a foolish, sophistic, and unqualified verbosity he identifies with his adversaries, the so-called Cornificians. Those are self-satisfied careerists who heap scorn on a literary education. Advocating brief study for material gain alone, they forgo the promise of *logos*.[34] "Some of our contemporaries," so John observes, "apparently pride themselves in being able to babble along garrulously without the benefit of this art. They regard it as useless, openly assail it, and glory in the fact that they have never studied it."[35] Far from basking in the apparent inutility of the verbal arts, John argues for a slow-going, synthetic, and instrumental literary education. His is a powerful critique based *in* instrumentality, and not one that is familiar nowadays, when we often see the term conflated with shallow designs and grasping personalities (akin to the Cornificians). For John, the situation was quite the reverse: an instrumental view of a liberal education is a bulwark against a tawdry transactionalism.

The thirteenth century saw the circulation of many other introductions to the liberal arts, including those of Robert Grosseteste's *De artibus liberalibus* (c. 1200), Bonaventure's *De reductione artium ad theologiam* (c. 1250), and as mentioned, Brunetto Latini's *Il Tesoretto* (ca. 1260). But *De ortu scientiarium* (c. 1250), composed by Robert Kilwardby, Parisian scholar and eventual archbishop of Canterbury, transmitted an outstanding account of the status of language and logic (which he calls *scientia scientiarium,* the science of sciences).[36] To that end, Kilwardby distinguishes between *possessio* and *usus,* employing the analogy of a cistern that is possessed by one person but is used by a community.[37] So it is with the subject of reasoning, which belongs to logic per se, having its own methods and concepts, while remaining "an instrument for all and support for all."[38] Instrumentality opens the container to many uses.

More technically, Kilwardby proposes a model not of opposition but of apposition, or what he calls *subalternatio.* Subalternation as a concept of logic derived from Aristotle's *Posterior Analytics* 1.7, 1.3, and 1.28, and applies to the conditions under which rational demonstrations can be carried out. For Aristotle, each science has a degree of autonomy within a larger structure of methodological relations. For example, arithmetic does not involve geometrical demonstrations, and geometry does not concern optics. The latter in these pairs is the "lower" subdiscipline in respect of the "higher" (the former in the pairs), from which are borrowed principles of demonstration. Medieval scholars variously elaborated on the idea in the translation and commentary tradition running through Gerard of Cremona, Robert Grosseteste, and John of Reading on down to Kilwardby.[39] We already saw something similar in the part–instrument distinction found in both Hugh and Gundissalinus. In Kilwardby, scientific subjects are diverse and independent. At the same time, subdisciplines stand in an interdependent relation of subalternation whereby one set of demonstrations effectively *descends* to another.[40] For example, mathematics is a discipline, concerning abstract measures with numbers, that need not itself engage in the sort of reasoning that belongs to geometry or astronomy. Geometry takes the measure of the earth, employing the resources of mathematics to that end; astronomy does the same for the stars, employing mathematical geometry. Likewise, further down the scale, the abstract resources of geometry (points, lines, and curves) and astronomy (magnitudes and motions) descend into increasingly applied subdisciplines like surveying. As Kilwardby observes, instruments like the astrolabe and sphere are subalternate to geometry in the same way that astronomy is subalternate to geometry.[41] Technical diagrams could be said to occupy a similar point in a cascading series, in the recognition of how abstractions are concretized there.

In his generous and synthetic vision of the arts disciplines, Kilwardby comes up with a capacious scheme that accounts for more than any single scholar could master. The point is that subalternation is a way of proceeding anywhere along the various routes (*viae*) to mastery. A starting point is grammar, which can be put to use in dialectic (reasoning by syllogism concerning universals) or rhetoric (speaking according to circumstances concerning singulars).[42] Grammar remains its own subject still, even as it descends into the subdisciplines. Such a division of the sciences is an idealization. It is also a useful heuristic, a demonstration of the ways the various *instrumenta* could be made to cohere: they are in his analysis transitive or mobile, one serving the other while remaining distinctive, leading him to see theoretical and practical sciences as reciprocal. One of Kilwardby's oft-quoted phrases is to the point: "The speculative sciences are practical and the practical sciences are speculative."[43]

Such lively didascalic discussions of the arts curriculum carried on into the following centuries, where the primacy of the word continued to reverberate outside of the university walls, as can be detected in the fourteenth-century case of Gower. His own account of the main branches of learning—*theorique, rhetorique,* and *practique*—represents an intriguing variation on previous anatomies of the arts. His classification is based on the work of the mid-thirteenth-century Florentine statesman, philosopher, and teacher of Dante, Brunetto Latini (d. 1294). Brunetto's compendious *Book of the Treasure* begins with a scheme inspired by Aristotle's *epistēmē, phronēsis,* and *technē,* but revised and rendered as *teorique, pratique,* and *logique.*[44] He himself departs from Aristotle in grouping ethical, political, mechanical, and rhetorical disciplines under one heading, that of *pratique.* Aristotle was clear that moral and mechanical skills are different kinds of activity, while Brunetto mixes them in what becomes a characteristically medieval manner. He also places facility with language and argument under *logique.* Gower, for his part, follows Brunetto up to a point, but clears an even larger space for the language arts. Whereas Brunetto had made rhetoric subsidiary, Gower promotes it to a ruling science itself. As Rita Copeland and Ineke Sluiter observe, "the category of rhetoric represents the power of the word, which, according to Gower, is the strongest instrument in human affairs."[45] Grammar and logic become subordinate and, in Gower's clear formulation, "serven bothe unto the speche" (*Confessio amantis* 7.1529). It is perhaps not surprising to find a poet ranking the arts of persuasion so highly, but as we have seen, he was not alone in advocating for the role of *speche* (*logos*). Nor was his notion of *service* unusual, if we recall Hugh's *usus* and Kilwardby's *subalternation*. Behind these ideas lies an instrumentalizing impulse and centuries of negotiation over different weightings in the *divisio scientiae*. Gower's view turns out to be akin to the other late classification systems where logic, language, and literature had a distinct utility within medieval Aristotelianism, augmenting the old *Organon*.

A final transitional example will help to reveal the legacy of the philological orientation we are identifying with medieval instrumentality, relating the concept more directly to literature and cultural criticism. Late in the fifteenth century, another Florentine humanist was attempting to reconcile the different branches of learning. Angelo Poliziano was preparing an introductory lecture for a course on Aristotle's *Prior Analytics,* part of the *Organon,* but before launching into a treatment of the syllogism, he decided to call out detractors who would accuse him of being no proper philosopher. Poliziano was a professor of poetics and rhetoric; he was no professional logician after all. His lecture attempts to dismantle such distinctions in a philosophical analysis, and he deploys the traditional idiom of instrumentality to resist disciplinary gate-keeping.[46] A fable serves as his opening defense, and explicitly so: "Stories are also—and just as often—philosophy's instruments" (*instrumenta philosophiae*).[47] *Stories,* not just *syllogisms,* are the rudiments of philosophy, and they are available to more people than trained philosophers. Poliziano goes on to relate a story his grandmother told him about fearsome female creatures, the so-called *lamias,* who by analogy are his carping opponents. The casual chauvinism evidently circulates sexist stereotypes, except that they are just as likely to be objects of the critique (an effective piece of grandmotherly wisdom). Which is it exactly? And why would such a question matter, or have any philosophic purchase? Poliziano shows that demotic discourse exemplified by "old wives' tales" has just as much value as dialectic, constituting able *instrumenta philosophiae.* Arguably, the move to foreground a *fabella* in *philosophia* draws on maligned matrilineal matter in a way that challenges antifeminist prejudices rather than propping them up. If that reading seems too forgiving, it is precisely the point that the means of persuasion and demonstration are to be opened up to more than one profession or class of writers and readers, allowing instead for philosophical disputes over the meanings of such meager things as fables.

However that may be, Poliziano's instruments of philosophy are broad-gauged and run through both liberal and mechanical arts; they undivide the *divisio scientiae*. Belonging to a period of great social and intellectual ferment whose central figures included Pico della Mirandola and Marsilio Ficino, Poliziano shared the view that higher education should be eclectic and extraordinarily expansive. "For Poliziano," observes Christopher Celenza, "philosophy is threefold: theoretical, practical, and rational. . . . Yet within those rubrics one finds references not only to metaphysics, physics, and ethics, but also to grammar, history, cooking, gladiatorial combat, carpentry, and tumbling, the latter art belonging to what Poliziano calls the 'jesting craftsmen' ('illos nugatorios artifices')."[48] Attempting to recover what he considered the highest ideals of classical learning, which are precisely the sort that can be studied only thanks to those like himself who engage in translation and commentary, Poliziano issues a critique of contemporary disciplinary norms that we too often fail to recognize as part of the humanist inheritance. It is one in which literature and literary criticism held pride of place among other *instrumenta*. Poliziano harks back to a time when philology had currency and authority: "Our age, knowing little about antiquity, has fenced the philologist in, within an exceedingly small circle. But among the ancients, once, this class of men had so much authority that philologists alone were the censors and critics of all writers."[49] He is making a larger claim about the commensurability and instrumentality of a literary education and the roving, transdisciplinary capabilities of the practicing *grammaticus, criticus,* or *literatus*. Those were the ones trained to read widely and with perspicacity. Poliziano recognized the risk of being treated as a dilettante, but nevertheless advocated for the language arts because they are fundamental to so many aspects of human endeavor. "Indeed," explains Poliziano, "the functions of the philologists are such that they examine and explain in detail every category of writers—poets, historians, orators, philosophers, medical doctors, and jurisconsults."[50] He also shared a basic assumption that the language arts are productive and instrumental to public life in general.

As others had argued, speech goes to form any and all higher concepts, and so it is no stretch to consider the language arts as essential to the formation of the educated intellect. Grammar, rhetoric, and poetics are critically instrumental, containing methods and devices with a capacity for different kinds of work. All *artes* are in fact so called because, as Thomas Aquinas once said, they "involve not only knowledge but also a work that is directly a product of reason itself: for example, producing a composition, syllogism or discourse, numbering, measuring, composing melodies, and reckoning the course of the stars."[51] The functions of the language arts would have been clear, and those who neglect the work of *grammaticus, criticus,* or *literatus* run against an emergent cross-disciplinary ideal that sustained centuries of considered opinion. In virtue of the eminently useful or instrumental interrelation among the disciplines, one has reason to reject a self-secluding, siloed disciplinarity that has, also perennially, sought to structure higher learning and teaching. Early scholars formulated a concept of *logos* as at once intellectual and applicable to endeavors outside of academic settings. They accepted that universities were set up to provide "a range of intellectual disciplines and techniques of direct community value."[52] Arts formed the basis of a broad, civic-minded, multidisciplinary liberal education, cultivating communicative strategies that can serve the larger polity. Accordingly, thinkers did not merely take an academic or antiquarian interest in the liberal arts. They typically prized the sources, techniques, and expertise of instrumental knowledge that obtained to public life.

MECHANICAL ARTS AND THE INVENTION
OF LITERATURE

The foregoing has focused on the production and circulation of discourse within scholarly contexts where communicative action was instrumentalized. If there is a risk in emphasizing the role of the grammar and rhetoric curriculum, it would be in possibly overlooking the concrete lingual and material conditions of knowledge production elsewhere, about which medieval thinkers also had something to say. Treatises on the liberal arts often included accounts of the mechanical arts, as already noted, illuminating further dimensions of instrumentality. Manual trades and technical inventions were considered applied disciplines whose craft tools and practical know-how also have a coherence, effectiveness, and eloquence. Both liberal and mechanical arts were viewed as productive disciplines in this regard. In fact, a variety of mechanical arts were congruent with, and sometimes even treated as constituent parts of, a larger intellectual apparatus.[53] To take a few notable examples, Gundissalinus and Gerard of Cremona both transmitted versions of al-Farabi's *scientia ingeniorum* (science of devices) and argued for the inventive and intelligible qualities of mathematical instruments that give physical form and function to doctrines handled elsewhere in a more conceptual way.[54] Kilwardby, in his nuanced account of the liberal and mechanical arts, described the phenomenon in another way: as a sort of subalternation descending from abstract theory to physical instruments. Under the practical sciences he also includes mechanics, which appropriately enough consists of seven disciplines.[55] He followed Hugh of St. Victor, who, for his part, had proposed that there are seven mechanical arts ("fabric making, armament, commerce, agriculture, hunting, medicine, and theatrics") resembling the seven arts of the *trivium* and *quadrivium*.[56] For these and other pliable thinkers, applied disciplines are granted considerable legitimacy in fields of knowledge production. Borrowing a conceit from Martianus Capella, Hugh calls the mechanical arts "seven handmaids which Mercury received in dowry from Philology, for every human activity is servant to eloquence wed to wisdom."[57] They are intimately related subdisciplines. It is perhaps an unequal marriage given the subordinate status of the mechanical arts in this configuration, but one in which technical objects and operations play no small part in the allegory of higher learning.

Yet going further, medieval thinkers are as likely to view grammatical signs and systems themselves as ingenious mechanisms—an original media technology consisting of a primary instrumentality. Philology produces the mechanical arts as nuptial gifts in a mythic foundation narrative, making sense of the degree to which applied sciences are reliant on language (and vice versa). But language is also a technical achievement. On a relatively literal view of *logos*, speech and writing are technologies that undergird human culture and communication. Gundissalinus already drew attention to the way the voice makes use of "natural instruments" (lips, teeth, tongue) and writing employs "artificial instruments" (hand, pen, parchment, ink). Among the latter must be included language as a contrived system laboriously learned, vocalized, inscribed, translated, disseminated, and put to so many creative purposes. The synthetic qualities of discourse would be all the more apparent in learned texts and traditions in Latin, Greek, and Arabic whose linguistic resources, as Karla Mallette observes, were the result of disciplines and techniques long circulating around the multilingual Mediterranean.[58] Considering the composition of any text, it can be hard to tell the liberal and mechanical arts apart. Didascalic writings that anatomize the arts are not just reporting on them; they are composed with them.

The point can be illustrated by returning to Gower's *Confessio amantis*. We saw how, in the seventh book, he upgrades *rhetorique* to a ruling science on par with *theorique* and *practique*, expressing the insight that *speche* (*logos*) is elemental. In the fourth book, Gower takes a more historical view of everyday language and craft knowledge, showing that words and writings are themselves crafted things so embedded in culture as to seem almost natural. They are instead artificial instruments for equipping knowledge production and circulation. He has readers understand that language makes various branches of learning possible because words, like farm implements and household utensils, are made to do the work. According to the Latin marginalia, Gower will describe those who first labored to discover the arts and sciences, "artes et sciencias primitus inuenerunt" (4.2377), using the verb *invenire* to suggest that the knowledge is a kind of wordy invention. He would expect readers to recall that rhetoricians compose speeches by employing the technique of *inventio,* the first of five classical parts of eloquence. Now invention is the basis of all the other technical *artes*. In English couplets, Gower goes on to describe those who, applying body and soul, developed agriculture and literary culture:

> Thus was non ydel of the tuo,
> That on the plogh hath undertake
> With labour which the hond hath take,
> That other tok to studie and muse,

As he which wolde noght refuse
The labour of hise wittes alle.
And in this wise it is befalle,
Of labour which that thei begunne
We be now tawht of that we kunne. (4.2382–90)

To Gower's way of thinking, the two kinds of labor developed in parallel. Ploughing and writing were often compared in the period, and that analogy evidently informs the verse: they represent two sorts of handiwork, of farmer and of scribe, and not incidentally the word *verse* itself comes from *versus* for furrow.[59] A significant feature of Gower's argument, then, is that labor and lore depend on writings that pass on the specialist skills required to cultivate wisdom. It is not a speculative notion; practical manuals would transmit knowledge in this way. The linkage was amplified in contemporary writings on farm equipment, as Lisa Cooper shows in her treatment of *On Husbondrie,* a fifteenth-century English translation of Palladius's *Opus agriculturae.* "Mak redy now vche needful instrument. / Let se the litel plough, the large also, / The londis forto enhaunce and vp to hent."[60] Concentrated attention on each subsequent instrument generates verse with a heightened literary and affective dimension, *furrowing* the pages with information about agronomy. Similarly, inventories of household tools appear in Cato, Varro, and Columella and in wordbooks such as Adam of Petit Pont's *De utensilibus,* Alexander Nequam's *De nominibus utensilium,* and John of Garland's *Dictionarius.* Those texts *on* tools become ready-to-hand rhetorical instruments, putting knowledge of useful things to work in the construction of useful texts.[61] And perhaps those texts will form the repositories of information out of which new compositions come, which brings us back to the rhetorical practice of *inventio.* What makes Gower's description of those who invented various labors so fascinating is its self-referentiality, grounding a poem, whose technical medium is language, in the language of technology.

As medieval rhetoricians and poets would have been the first to recognize, making anything new requires an *inventory* of preexistent elements out of which innovative works can be composed. That is one of the "paradoxes of the new" that Patricia Ingham has investigated in a range of medieval literature.[62] Isidore of Seville helpfully explains by drawing on the inventory of etymons found embedded in words: "If we reconsider the origin of the word, what else does it sound like if not that 'to invent' [*invenire*] is to 'come upon' [*in + venire*] that which is sought for?"[63] The very term "invent" *enacts* the idea of finding meaningful things in language.

Gower's *Confessio* is equally illuminating with respect to the way language discovers, assembles, distributes, and perpetuates meaning. In a section that relies heavily on Hugh's *Didascalicon,* Gower catalogues those who first developed letters and language arts: Noah devising Hebrew and the seven liberal arts; Cadmus coming up with Greek; Carmentis developing the Latin alphabet; Herodotus inventing meter, rhyme, and cadence; and Aristarchus together with Donatus and Didymus originating the "ferste reule of scole" (*Confessio amantis* 4.2642)—grammar. Cicero is said to have developed rhetoric: "Hou that men schal the words pike / After the forme of eloquence, / Which is, men sein, a gret prudence" (4.2651–53). Historical accuracy is surely not the aim in the sketchy legendary matter; the catalogue of famous originators rather reads as a heuristic in the service of a broader idea about the production of knowledge. It recalls that the poet and reader are beholden, in the composition of any document, to the cultivation of linguistic resources. The same goes for translators who labored over scripture and every other kind of lore to sustain a large textual inheritance:

> . . . out of Hebreu
> Jerom, which the langage kneu,
> The Bible, in which the Lawe is closed,
> Into Latin he hath transposed;
> And many an other writere ek
> Out of Caldee, Arabe, and Grek
> With gret labour the bokes wise
> Translateden. And otherwise
> The Latins of hemself also
> Here studie at thilke time so
> With gret travaile of scole toke
> In sondri forme for to boke,
> That we mai take here evidences

> Upon the lore of the sciences,
> Of craftes bothe and of clergie;
> Among the whiche in poesie
> To the lovers Ovide wrot
> And tawhte, if love be to hot,
> In what manere it scholde akiele. (4.2653–71)

The verbal force of "for to boke" indicates that texts are technological innovations, merging mechanical and liberal arts in any act of communication. Those communications are as technically ambitious as they are transcultural: Hebreu, Caldee, Arabe, Grek, and Latin. They are also interdisciplinary. In this scenario, book-learning results from various material affordances, manual acts, and cultural sources as much as from abstract intellectual inquiries, requiring everything from alphabets to authors who can devote themselves to agronomy or amatory arts. Gower does not mention all the contributing causes of culture and communicative media, though his references to agriculture, religion, and rhetoric embrace a large spectrum of technology.

So many practical disciplines go into documenting human knowledge, and they include productive acts that result in the very documentary record of those disciplines. Knowledge dissemination requires numerous and complex tasks, tools, and resources, which are then made into works with a capacity for yet more work. Just taking manuscript manufacture, for example, the making of books involves breeders, skinners, parchmenters, scribes, limners, and annotators whose labor involves everything from selecting animal hides to transforming the membrane into a consistent writing surface, copying and annotating, drawing and painting, gathering leaves into quires, and binding a whole assemblage. This rapid description is awfully general and incomplete, but allusions to materials and practical know-how suggest that the making of books is a phenomenal technical accomplishment. In this straightforward respect, the mechanical and liberal arts are reciprocal inventions recalling something Cassiodorus wrote centuries earlier when supposing that *artes liberales* comes from *liber,* or book.[64] An inescapable instrumental condition frames and makes possible all theoretical and practical endeavors after all.

Conclusion

Toward a Critical Instrumentality

Instrumentality has been a perennial subject of debate among those committed to higher education in an ever increasingly commercialized and technologized society. The debasement of learning to something transactional must continue to concern anyone who wishes to resist the reduction of life to economic efficiency under capitalism. One of the main points of this book, however, is that recent worries about the costs of an instrumentalizing modernity are subsequent moments in an intellectual history stretching back to premodern articulations of matter and method. And as we have observed, medieval thinkers expressed a sophistication that has largely disappeared in treatments of instrumentality today. An instrument once encompassed anything from sense faculties to mental concepts, manual objects, information systems, and scholarly disciplines, even characterizing a course of study in the liberal arts. All the while, instrumentality could stand opposed to mere technicity or transactionalism.

Where has that subtlety and conviction gone? I will conclude by suggesting that prominent modern critics, unable to shake the exigency of the conceit even as they engage in vigorous critiques of instrumentality, have failed to attend to the pragmatic consequences of their most trenchant theories. Several objections have been leveled in contemporary theoretical discussions, and they help explain why professional scholars may feel allergic to talk of instrumentalism. I share a distrust of crudely utilitarian approaches to higher learning. An emergent technocracy that threatens to swallow education whole must be resisted. But criticism of this sort does not always perceive its own principles of utility; critique is not everywhere sensitive to the conditions under which it operates and becomes instrumental within scholarly and social spheres. Is any critical intervention possible without wielding useful tools and techniques, even if aiming at a radical retooling? How long can practical reason be suspended in the pursuit of alternative paradigms and strategies? Does instrumentality not remain a deep property of thinking and writing in the first place?

Perhaps most recognizable in media and technology studies is the theory that the authentic meaning of technology is lost in incriminating use and practical relations; one is instead meant to meditate on the character of tools apart from anything so profane as a capacity for work. So runs Martin Heidegger's famous tool analysis: a hammer put to use is ready-to-hand, while one that is broken becomes present-to-hand and escapes the reduction to mere equipment.[1] What he calls "equipmentality" renders a tool "always is *in terms of* its belonging to other things: ink-stand, pen, ink, paper, blotting pad, table, lamp, furniture, windows, doors, room."[2] Yet the Heideggerian search for a primal and universal truth makes the object of inquiry nearly vanish from history. In the desire for an enigmatic and sublime presencing of technology, the philosopher gives up a useful account of everyday practices that apparently don't rank philosophically. A practical and historical orientation such as that modeled by medieval thinkers, supporting a more granular technographic analysis, goes some way toward showing how technical objects matter: shaping routines, framing views, directing energies, representing possible ideas, forms, and relations. The scene of writing is no less instrument-bound. What, we may ask, do Heidegger's own ink stand, pen, ink, paper, blotting pad, table, and so forth bring to philosophical work? How are the technical requirements of philosophical composition manifest in his own writing about the *instrumenta scribae* (to recall Isidore of Seville)? Theory appears to be reliant on craft tools and technique, writing supports, manual and graphic instruments, and even just iterable marks on the page. "Tool" is an attenuated term to use in this context, but by thinking—even experimentally—of a written work as one telling example, we can hopefully enter into more capacious dimensions of the question. A welcome stimulus for this inquiry is found in Gilbert Simondon's *On the Mode of Existence of Technical Objects,* which attempts to describe the cultural components and historicity of technical activity. Technicity is often reduced to the minimal procedural functioning of a machine, rather than being seen as an integral part of the social fabric, which explains why tools get segregated from aesthetic culture.[3] Simondon argues for an understanding of the technical imagination that generates a cultural milieu in the first place. Bernard Stiegler's *Technics and Time* also thinks through and beyond the way "philosophy has repressed technics as an object of thought."[4] Radicalizing Simondon's thesis, Stiegler posits that humanity and technicity are coeval as registered in what he calls "the instrumental condition" of being human.[5] Perhaps medieval sources had already intimated that so many aspects of everyday life are constitutively technical and ineluctably instrumental.

A potentially more powerful contemporary critique stems from ecopolitical analyses of an aggressive postindustrial and technocratic modernity, where the instrumental condition is a source of well-founded worries about the fate of planetary life.[6] The present epoch is poised at the edge of imminent global catastrophe owing to dehumanizing routines and alienating machines, ever more efficient extraction of natural resources, technological surveillance, and exploitation of labor. I share the general critique of capitalist or technocratic instrumentalities. Yet, in this context, we may carefully avoid the kind of technological determinism that too quickly identifies the sources of trouble with the machinery of the moment and excludes alternative implementations.[7] It is from a historical standpoint that ecocritique might gain additional force, revealing foreclosed possibilities to instrumentality before the full onslaught of global capitalism and technoscience. As this book indicates, there is a chance to reimagine instruments as not just maximally efficient but also ornamental, intellectual, devotional, cosmological, imaginative, and more, all considered within the ambit of a tool's capacity for work. An instrument can model more than one future after all. A physical globe, for instance, is one early (medieval) and enduring technology that accesses and represents planetary scales and processes, and that has at different times played a role in astronomical theory, pedagogy, cartography, colonialism, and globalization. It is surely an ambivalent object, one as capable of enabling slow violence as of making the sources and effects of violence visible (e.g., mapping the unequal impacts of climate change as an index of environmental racism).[8] A globe, in short, argues against technological determinism. Of course, there must remain a healthy suspicion of technologies in a world replete with ever new contrivances, so long as we do not mistake the situation for one where the choice is to down our tools. *Instrumentality* supposes that most technical ensembles have multiple overlapping and variable affordances. It is for that reason that it is crucial to maintain the cross-disciplinary and technical capacities required to comprehend the liabilities and possibilities of media and technology. How else is there to recognize the menace of transnational capital, digital surveillance, toxic waste, and exploitive labor systems in the digital age? Nor is the aim to elevate a positivist paradigm above others. The book has shown that the more competent the technical instrument, the more it involves mediation and interpretation. Humanistic scholarship itself relies on so many fine-tuned instruments.

Another sort of suspicion is found in plaints against "instrumental reason" that have their sources in the thought of Max Weber, Antonio Gramsci, and the Frankfurt School. An essential problem was the threatened dominance of the state apparatus and monopolistic industrial powers. Gramsci argued that schools were "instruments" of hegemony.[9] Theodor W. Adorno feared a completely administered society "organized according to the dictates of instrumental reason and the prerogatives of managerial experts in the name of economic growth and technological progress."[10] Bureaucratic rationalization was threatening to compel assent, depersonalize, routinize, and marginalize. The same phraseology would become central to Charles Taylor's account of modern selfhood formed around a shallow, secular individualism. For Taylor, an instrumentalist self-conception now counts among the limited and lamentable repertoires of identity, eclipsing an ancient and medieval moral order.[11] A comparable idea has defined the humanistic enterprise more locally against the impersonal metrics and audits of an increasingly administered university. Stanley Fish's protestations against the "instrumental defence of the humanities" have been among the most audible. "The distinctiveness of the humanities and liberal arts education rests on their inutility," writes Fish, "on their fostering a mode of thought that does not lead (at least by design) to the 'practical' solution of real-world problems but to deeper understanding of why they are problems in the first place and why they may never be resolved."[12] It is a nearly reflexive stance of contemporary literary humanists to array themselves against transactional or utilitarian managerial imperatives. Such critiques of instrumentalism surely strike a sympathetic chord, but ultimately overextend themselves and commit critics to some unexpected conclusions about early and late intellectual formations in which they participate. For, if Gramsci is right about the role of schools in the practical organization of society, then the goal must be to augment and improve rather than abandon them, as he himself was aware.[13] Taylor, for his part, underestimates the overt instrumentality involved in early philosophical, scientific, and religious work, preferring a universalizing version of medieval mysticism and quietism of his own manufacture; and he concedes too much to the novelty of the instrumentalist stance in modernity. Fish's turn against "utility" neglects how humanists have always been beholden to instrumental systems and rationales, assessing and credentializing students, regulating and rewarding academic labor, and authorizing research methods that legitimate the intellectual detachment to which he would adhere. Actually, Fish's energetic pronouncements in the press show an admirable ethical interest in civic engagement and the commons, going well beyond the bounds of what, on his own account, is relevant to the literary scholar. A longer history of the arts curricula shows how instrumental rationality has been part of their self-definition from the start, which is something the critics do not seem to know.

Hopefully we can by now see that the question has never been whether the liberal arts are instrumental, but in what ways. The choice is not between engagement and disengagement, but a matter of which interests to pursue within the inevitable institutional constraints. What can be witnessed in a recent pattern of critiquing instrumentalist orientations is really a complaint against *specific* transactional means and extractive ends of scholarship, and against adherence to vacuous notions of bureaucratic efficiencies and so-called "excellence," *not* a wholesale repudiation of instrumentality. The critics just have better purposes in mind for, say, the study of literature, philosophy, or history. To the doubters and cynics, any alleged scholarly detachment will appear covertly political anyway (i.e., *useful* or *advantageous* to an imagined enemy), and that is to register at least a partial truth.

Academicians across the arts and sciences pursue working understandings, and occasionally they create new knowledge (historical, legal, epistemological, and aesthetic) that becomes the basis for reasoned social or political changes. My point is not to rehearse a weak claim in the oft-expressed view that a liberal education presents so many chances to develop transferrable skills. That sort of positive instrumentalism—"we teach students *how* to think, not *what* to think"—has become fairly widely accepted, but can seem disingenuous. For one thing, methodological rigor alone cannot adequately inform students and make up an educated citizenry if it amounts to an empty proceduralism. The idea is especially wide of the mark when phrases like "critical thinking" and "doing your own research" have made it into the populist vernacular to justify all sorts of false conspiracist beliefs and illiberal policies nowadays. Nor does isolating the *how* from the *what* describe the substantive learning objectives of individual courses one finds in universities today. Biology undergraduates are not merely learning how to learn; they are acquainted with the useful facts of morphology, anatomy, life cycles, ecologies, and so on. Courses on the Jewish Holocaust or feminist media studies are not simply grist for the mill of "critical thinking," but venues in which students come to grips with the reality of past events, bear witness to traumas, and form truer ideas of themselves as social and historical actors. Nor are those kinds of knowledge reducible to vocational ends, even as they most definitely have personal and political consequences. Louis Menand is one of the few to make the point explicit: "Knowledge just *is* instrumental; it puts us into a different relationship with the world."[14] More recently, Eric Hayot has laid out the case for the "social effectiveness of humanist work in such realms as the law, social formations, self-conceptualization, institution-building, or aesthetic production." He concludes by issuing an important challenge: "for the humanities to become responsible for their practice."

Historians have taught us that romanticized notions of scholarly aloofness could never have created the conditions in which a disinterested professoriate was possible in the first place.[15] The university is a legal and corporate apparatus that has always been a highly interested and purposive entity, with the very term *universitas* signifying a working guild or corporation. That is nowhere more obvious than in the Middle Ages, says Alan B. Cobban: "The roots of the medieval university phenomenon were formed in utilitarian soil. Europe's earliest universities were institutional responses to the need to harness the expanding intellectual forces of the eleventh and twelfth centuries to the ecclesiastical, governmental and professional requirements of society."[16] That helps explain a preoccupation then with the "instruments of philosophy." Yet academic attainments and appointments continue to be instrumentalizing in new and diverse ways, not least in qualifying graduates for a changing job market. At least some of the time academics also reproduce themselves, and how they do so is something for which they often feel great responsibility. Frank Donoghue writes: "Many would prefer to think of their intellectual labour as something that stands outside ordinary definitions of work. This is a dangerous assumption."[17] If anything, endorsements of an otherworldly humanities are liable to serve the austerity policies of an opportunistic educational establishment, concealing a dependence on casualized labor, legitimizing procrustean business strategies, and maintaining a preference for short-term outcomes. The arts disciplines will remain forever on the back foot so long as they mystify or misidentify the actual working conditions. It is precisely because of the dangers of a thoroughly administered educational apparatus, preferring abstract metrics over sense-making, that the liberal arts must strive to enlarge the language and ethos of instrumentality.

Scholarship has developed sophisticated rationales that help illuminate the forces and effects of higher learning, avoiding the insularity and defeatism of anti-instrumentality. One could riff on major critical themes from the last several decades speaking to the way all intellectual activity has a share in the social and political. What models of education might follow from established ideas that knowledge is conditioned by ideology, or constituted by speech acts, or dependent on actor networks, or subject to media effects? How could we best account for the generative effects of reasoning, interpreting, composing, and critiquing together?

Answers vary, but seem nonetheless committed to the agency and utility of the arts disciplines, whether manifest in social critique, political praxis, or some other transformative purpose. Many agree that a liberal education is profoundly *constructive*. We hear that humanist scholarship is a "form of reason" as serviceable as any methods associated with the physical sciences: "Humanist reason builds models of causation, of meaning, and of life, that aim to explain pasts, understand presents, and imagine futures."[18] Or that, as John Guillory writes, the "mode of rationality" of the literary humanities counts as *techné*, in which "the study of literature is the site for the development of modes of cognition specific to language use" that can unlock practical potentialities needed "to respond to, or to create, highly wrought instances of speech."[19] Others like Rita Felski see literary study among other arts as functioning to cultivate affective bonds, negotiate competing desires and interests, and teach "generous interpretations."[20] Alongside epistemological, rhetorical, and ethical uses we find overtly political and progressive ones. Paulo Freire long ago spoke of a "pedagogy of the oppressed" that serves as a life-affirming "instrument for their critical discovery." A dialogical model of cultural action is presented as "an instrument for superseding the dominant alienated and alienating culture."[21] Adopting the phraseology of Freire, bell hooks has argued that a liberal education is a positive praxis: while academic theory can be an "instrument of domination," it may "also contain important ideas, thoughts, visions, that could, if used differently, serve a healing, liberatory function."[22] Freire and hooks both present recuperative strategies that have become important stimuli among scholars working to establish alternative reparative and coalitional politics. They are in search of better principles of utility. Sarah Ahmed's *What's the Use?* is one sustained negotiation with instrumentality in academic institutions where minds and bodies are asked to undertake useful work. Building on hooks, Ahmed proposes a new expedient: "Our task in challenging instrumental rationality is to make use of other uses of use."[23] Guided by the oft-cited dictum of Audre Lorde ("the master's tools will never dismantle the master's house"), she might be supposed to abandon anything like instrumentality, but even demolition projects require the right tools, and Ahmed has been most successful in putting new ones into circulation—including eminently practical means for building support systems among the marginalized. Affiliated queer, antiracist, and decolonial initiatives express and embody an urgent purposiveness within contemporary arts disciplines. Many humanist scholars are attending to the enduring wrongs of imperialism, systemic racism, and white supremacy, including those perpetuated in and through the educational apparatus. Indigenization is the most recent instrumental imperative to reinvigorate academic institutions. My university's Indigenous Plan, promoting the calls to action of the Canadian Truth and Reconciliation Commission, acknowledges the political exigencies and opportunities for Indigenous knowledge dissemination and community-building on campus.[24] Urging disciplines to mobilize around a common cause, implementing hard-won historical understandings of past and present injuries, the document charts a way toward what can be called forms of *critical instrumentality*. Of course, what is decisive in these novel formulations of instrumental reason is parsing the difference between, on the one hand, implements or applications that are narrowly transactional and technocratic, and on the other, those that are intellectually expansive and socially responsible.

One benefit of so reasoning is to generate ideas of the university that are a powerful match for capitalist instrumentalizations like the prevailing corporate ethos that would reduce higher education to replicating the workforce or sustaining the world-ruining regime of neoliberal commerce. Another is to offset a prevailing bias toward the so-called "hard sciences," not by further separating the humanities from STEM subjects (science, technology, engineering, and mathematics), but by showing that a liberal education has long included many kinds of inquiry that are critically instrumental. We can return art and agency to instrumentality. Humanist scholars usefully address a large spectrum of human experience and intelligence, amplifying diverse voices and visions; teaching and revitalizing languages; and recollecting histories of cross-cultural exchange, peace and conflict, sovereignty and subjection, law and social disorder, nature and culture, religion, race, sex, and gender (to name only a few subjects); preparing generations of students to enter public life and participate in a diverse global polity. Academia retains the special capacity to forge instruments of diverse arts, whose reserves of historical information, imagination, invention, and analysis can shape a more just and livable future. Ongoing issues such as climate change, pandemic disease, viral disinformation, gender-based violence, colonial dispossession, and racial inequalities demand transdisciplinary solutions. Linear, statistical thinking is never enough; complex problems cannot be repaired without attending to a dense web of social relations, historical conditions, and political responsibilities in which human action is bound up. The present moment requires more than the empty verbiage of "enterprise" and "innovation" can deliver in the fields of science and technology, and a recognition that many fine instruments of logic, language, and literary analysis will continue to be indispensable in generating and sustaining positive social change.

Notes

Introduction

1. Paris, Bibliothèque nationale de France [BnF], MS Latin 7294, fol. 31r: "Totius astrologie speculationis radix et fundamentum eiusque prolixitatis immensitas subtilitatisque inexhausta profunditas ex hiis que per visum percepta sunt non sine congruis instrumentis sumpsit exordium. Certum nanque apud omnes huius discipline professores habetur instrumentis astrologie sublatis nullam motus celeste circuli corporumque celestium cognoscendi viam derelinquit. Horum autem instrumentorum licet magna sit numerositas. In hoc tamen generaliter videntur esse concordes, quod omnes motum celi imitantes et ad ipsius similitudinem velut ad exemplar" (trans. slightly modified from Richard Lorch, "The *sphera solida* and Related Instruments," *Centaurus* 24 [1980]: 155–56).

2. It survives in over thirty manuscript copies and three sixteenth-century imprints, enjoying wide circulation, and the work is occasionally attributed to Accursius de Parma. See the bibliography in Lynn Thorndike and Pearl Kibre, *A Catalogue of Incipits of Mediaeval Scientific Writings in Latin* (Cambridge, Mass.: Medieval Academy of America, 1963), 1576. For a discussion of authorship and the content of the treatise, see Kathrin Chlench, "'Sphera solida'—A Treatise on the Celestial Globe and Its Significance in Late Medieval Astronomy," *Globe Studies* 57–58 (2011): 70–80; and Elly Dekker, *Illustrating the Phaenomena: Celestial Cartography in Antiquity and the Middle Ages* (Oxford: Oxford University Press, 2012), 341–43. In omnibus collections of instrument treatises and associated theories (e.g., BnF, MS Latin 7294; and London, British Library, Royal MS 12 C XVII), Harlebeke's work on the sphere solid is bound together with texts dedicated to quadrants old and new, cylinder, torquetum, and astrolabe.

3. John of Lignières: "Quia nobilissima scientia astronomie non potest bene sciri sine instrumentis debitis, propter quod necessarium fuit componere instrumenta in ea. Composuerunt ea propter antiqui multa diversa instrumenta ut sunt astrolabium et saphea, cum quibus sciuntur plura tam de tempore quam de motu" ("A Merton College Equatorium: Text, Translation, and Commentary," ed. and trans. Seb Falk, *SCIAMVS* 17 [2016]: 136–37). John of Lignières's prologue was repurposed in later equatorium treatises, and for another version, see appendix 2 of *The Equatorie of the Planetis: Edited from Peterhouse MS 75.1*, ed. D. J. Price (Cambridge: Cambridge University Press, 1955), 188. A modeling and calculating instrument, the equatorium consists of carefully calibrated plates and dials for expressing the stations, directions, and retrogradations of the planets.

4. Geoffrey Chaucer, *A Treatise on the Astrolabe*, ed. Sigmund Eisner, in *The Prose Treatises*, Variorum Edition of the Works of Geoffrey Chaucer 6/1 (Norman, Okla.: University of Oklahoma Press, 2002), 105. All subsequent citations to the treatise will appear in parentheses.

5. Francis Bacon, *Novum Organum* 1.2: "Nec manus nuda, nec intellectus sibi permissus, multum valet; instrumentis et auxiliis res perficitur" (ed. Thomas Fowler [Oxford: Clarendon, 1889], 192; trans. adapted from *The Works of Francis Bacon*, vol. 4, *Translation of the Philosophical Works 1*, ed. and trans. James Spedding, Robert Leslie Ellis, and Douglas Denon Heath [London, 1857–1874], 47; cited in Thomas L. Hankins and Robert J. Silverman, *Instruments and the Imagination* [Princeton, N.J.: Princeton University Press, 1995], 3).

6. *Oxford English Dictionary*, s.v. instrument; *Dictionary of Medieval Latin from British Sources*, s.v. instrumentum.

7. Isidore, *Etymologies* 5.25.26–28: "Instrumentum est unde aliquid construimus, ut cultrus, calamus, ascia. Instructum, quod per instrumentum efficitur, ut baculus, codex, tabula. Vsus, quem in re instructa utimur, ut in baculo innitere, in codice legere, in tabula ludere" (*Isidori Hispalensis episcopi Etymologiarum sive originum libri XX*, 2 vols. ed. W. M. Lindsay [Oxford: Clarendon, 1911]; also at https://penelope.uchicago.edu /Thayer/E/Roman/Texts/Isidore/home.html; trans. from *Etymologies of Isidore of Seville*, ed. and trans. Stephen A. Barney, W. J. Lewis, J. A. Beach, and Oliver Berghof [Cambridge: Cambridge University Press, 2006]).

8. Isidore, *Etymologies* 6.14.3: "Instrumenta scribae calamus et pinna." Bartholomeaus Anglicus notes that sometimes angels are painted bearing

inkhorns and other "instrumentis of writters," since by their actions we know the will of God, as indicated in *On the Properties of Things* 2.4, in *On the Properties of Things: John Trevisa's Translation of De proprietatibus rerum of Bartholomaeus Anglicus: A Critical Text,* ed. M. C. Seymour, vol. 1 (Oxford: Clarendon, 1975–88), 65.

9. Isidore, *Etymologies* 2.26: "Sequuntur Aristotelis categoriae, quae Latine praedicamenta dicuntur: quibus per varias significationes omnis sermo conclusus est. Instrumenta categoriarum sunt tria." The same language is found in John of Salisbury, *Metalogicon* 151, 155.

10. Jerome Taylor, trans., *The Didascalicon of Hugh of St. Victor: A Medieval Guide to the Arts* (New York: Columbia University Press, 1991), 88; Charles Henry Buttimer, ed., *Hugonus de Sancto Victore Didascalicon, De Studio Legendi: A Critical Text* (Washington, D.C.: Catholic University of America Press, 1939), 55.

11. Bruce L. Venarde, ed. and trans., *The Rule of Saint Benedict,* Dumbarton Oaks Medieval Library (Cambridge, Mass.: Harvard University Press, 2011), 32 and 35–37; cited in Lisa H. Cooper, *Artisans and Narrative Craft in Late Medieval England* (Cambridge: Cambridge University Press, 2011), 113. A habit of calling the sacred scriptures themselves *instrumenta* can be witnessed early and late by the likes of Tertullian, Aldhelm, Bede, and John of Salisbury. Tertullian speaks of the two testaments as instruments ("alterius instrumenti vel . . . testamenti") in *Against Marcion* 4, as discussed in Wolfram Kinzig, "Καινὴ διαθήκη: The Title of the New Testament in the Second and Third Centuries," *Journal of Theological Studies* 45, no. 2 (1994): 539–40. In *Apology* 18, Tertullian states that God revealed himself in books ("instrumentum litteraturae"), a phrase that can be compared to his description of pagan literature in *De Spectaculis* 5 ("de instrumentis ethnicalium litterarum"); *Apology, De Spectaculis, Minucius Felix: Octavius,* trans. T. R. Glover and Gerald H. Rendall; Loeb Classical Library 250 (Cambridge, Mass.: Harvard University Press, 1931), 88–89 and 242–43. See also *Dictionary of Medieval Latin from British Sources,* s.v. *instrumentum, def. 4.*

12. See Lisa H. Cooper, "The Poetics of Practicality," in *Oxford Twenty-First Century Approaches to Literature: Middle English,* ed. Paul Strohm (Oxford: Oxford University Press, 2007), 493. Household anthologies containing conduct texts, saints' lives, exemplary tales, and so on were directed at shaping the affective and spiritual lives of readers, serving as

what Myra Seaman calls "instruments of moral correction"; for a careful reading of one such anthology see *Objects of Affection: The Book and the Household in Late Medieval England* (Manchester: Manchester University Press, 2021), 165.

13. See, for instance, the recent discussion of personification allegory in Katherine Breen, *Machines of the Mind: Personification in Medieval Literature* (Chicago: University of Chicago Press, 2021).

14. The winning combination of use and enjoyment in literary work was extolled in Horace's *Ars Poetica,* but the idea also informed other activities. As Valerie Allen recalled to me in conversation, Old English *brucan* combined the senses of use and enjoyment, just as did the late Latinate blend *usufruct* (literally "use-enjoyment"), a legal idea granting a temporary right and advantage.

15. Cicero, *On the Nature of the Gods, Academics,* trans. H. Rackham, Loeb Classical Library 268 (Cambridge, Mass.: Harvard University Press, 1933), 508–9; Seneca, *Epistles,* trans. Richard M. Gummere, 3 vols., Loeb Classical Library 75–77 (Cambridge, Mass.: Harvard University Press, 1917–25), 2:360–61, 3:62–63; Boethius, *Theological Tractates, The Consolation of Philosophy,* trans. H. F. Stewart, E. K. Rand, and S. J. Tester, Loeb Classical Library 74 (Cambridge, Mass.: Harvard University Press, 1973), 416–17; Robert Grosseteste, *De artibus liberalibus,* ed. and trans. Sigbjørn Olsen Sønnesyn in *The Scientific Works of Robert Grosseteste,* vol. 1, ed. Giles E. M. Gasper et al. (Oxford: Oxford University Press, 2019), 74–75.

16. *Dictionary of Medieval Latin from British Sources,* s.v. *organum.*

17. Bartholomaeus Anglicus, *On the Properties of Things* 4.1 (Seymour, 1:129).

18. Aristotle, *De anima* 432a1–4, on which see Ronald Polansky, *Aristotle's "De Anima": A Critical Commentary* (Cambridge: Cambridge University Press, 2010), 496–97.

19. Bartholomaeus Anglicus, *On the Properties of Things,* 5.28 (Seymour, 1:222).

20. *Middle English Dictionary,* s.v. *ournement; Dictionary of Medieval Latin from British Sources,* s.v. *ornamentum.*

21. Roger Bacon, *Fr. Rogeri Bacon Opera quaedam hactenus inedita,* ed. J. S. Brewer (1859; repr. Cambridge: Cambridge University Press, 2012), 532.

22. Bartholomaeus Anglicus, *On the Properties of Things* 5.28 (Seymour, 1:223).

23. Besides the practice of finger-reckoning, there was the "computus hand" and the "music hand" (the one embodying a sort of onboard human abacus, and the other a musical scale arrangement). See John Haines, "The

Visualization of Music in the Middle Ages: Three Case Studies," in *The Visualization of Knowledge in Medieval and Early Modern Europe*, ed. Marcia Kupfer et al. (Turnhout, Belgium: Brepols, 2020), 331–34; Carol Berger, "The Hand and the Art of Memory," *Musica Disciplina* 35 (1981): 87–120.

24. See Cooper, *Artisans and Narrative Craft*, 190.

25. Jentery Sayers, "Technology," in *Keywords for American Cultural Studies*, ed. Bruce Burgett and Glenn Hendler (New York: New York University Press, 2014), 237. The rise of instrumental rationality has occasioned vigorous critiques leading to the further narrowing and pejoration of the term, associating it with shallow thinking and worse; for an account of the complex history, see Darrow Schecter, *The Critique of Instrumental Reason from Weber to Habermas* (New York: Continuum, 2010).

26. Bernard Stiegler, *Technics and Time*, vol. 1, *The Fault of Epimetheus*, trans. Richard Beardsworth and George Collins (Stanford, Calif.: Stanford University Press, 1994), 1. As Eric Schatzberg reminds readers, however, "there is no direct path from ancient ideas about techne and our modern concept of technology" (*Technology: Critical History of a Concept* [Chicago: University of Chicago Press, 2018], 16).

27. William Clark, *Academic Charisma and the Origins of the Research University* (Chicago: University of Chicago Press, 2006), 6.

28. The idea of a mathematical "scientific instrument" as contrasted with the tools of trade, commerce, engineering, and music is a nineteenth-century conceit, on which see Deborah Jean Warner, "What Is a Scientific Instrument, When Did It Become One, and Why?," *British Journal for the History of Science* 23, no. 1 (1990): 83–93, which was followed up by Lisa Taub, "What Is a Scientific Instrument, Now?," *Journal of the History of Collections* 31, no. 3 (2019): 453–67.

29. That the category of "the literary" as sometimes bound up with a sense of what is useful, instructive, or practicable is addressed by contributors to the recent essay collection *The Medieval Literary: Beyond Form*, ed. Robert J. Meyer Lee and Catherine Sanok (Boydell & Brewer, 2019), 15–84 (part 1, "Instrumental Forms"). Claire M. Waters observes in her treatment of religious verse: "In many medieval texts, the boundary between form and use, between the literary and the utilitarian, is not so much fluid as undetectable" (19).

30. The bibliography is vast and growing, but a few studies that have been influential in my thinking are: Wendy Chun, *Programmed Visions: Software*

and Memory (Cambridge, Mass.: MIT Press, 2011); Matthew Kirschenbaum, *Mechanisms: New Media and the Forensic Imagination* (Cambridge, Mass.: MIT Press, 2012); Alexander Galloway, *The Interface Effect* (Cambridge: Polity, 2012); Lori Emerson, *Reading Writing Interfaces: From the Digital to the Bookbound* (Minneapolis: University of Minnesota Press, 2014); Jussi Parikka, *Insect Media: An Archaeology of Animals and Technology* (Minneapolis: University of Minnesota Press, 2010) and *A Geology of Media* (Minneapolis: University of Minnesota Press, 2015). Yet we should be careful not to collapse the differences between old and new media. A medieval scientific apparatus embodies unwired, analog-era technology, and brings new contexts to bear on debates in the digital humanities and media studies.

31. Major sources include Donna Haraway, *Simians, Cyborgs and Women: The Reinvention of Nature* (New York: Routledge, 1991); Andy Clark, *Natural-Born Cyborgs: Minds, Technologies, and the Future of Human Intelligence* (Don Mills, Ont.: Oxford University Press Canada, 2004); Bruno Latour, *Aramis, or the Love of Technology*, trans. Catherine Porter (Cambridge, Mass.: Harvard University Press, 1996); Silvia Benso, *The Face of Things: A Different Side of Ethics* (Albany: State University of New York Press, 2000); Karen Barad, *Meeting the Universe Halfway: Quantum Physics and the Entanglement of Matter and Meaning* (Durham, N.C.: Duke University Press, 2007); Stacy Alaimo, *Bodily Natures: Science, Environment, and the Material Self* (Bloomington: Indiana University Press, 2010); Jane Bennett, *Vibrant Matter: A Political Ecology of Things* (Durham, N.C.: Duke University Press, 2010); Tim Ingold, *Being Alive: Essays on Movement, Knowledge and Description* (New York: Routledge, 2011).

32. Elizabeth Grosz, *Time Travels: Feminism, Nature, Power* (Crows Nest, Australia: Allen & Unwin, 2005), 146–47.

33. Stiegler, *Technics and Time,* 1:193, 206.

34. Barad, *Meeting the Universe Halfway,* 169.

35. See Brian Rotman, *Mathematics as Sign: Writing, Imagining, Counting* (Stanford, Calif.: Stanford University Press, 2000) and *Becoming Beside Ourselves: The Alphabet, Ghosts, and Distributed Human Being* (Durham, N.C.: Duke University Press, 2008); Catarina Dutilh Novaes, *Formal Languages in Logic: A Philosophical and Cognitive Analysis* (Cambridge: Cambridge University Press, 2012).

36. Anna Kornbluh, *The Order of Forms: Realism, Formalism, and Social Space* (Chicago: University of Chicago Press, 2019), 7–10.

37. Steven Connor, *Living by Numbers: In Defence of Quantity* (London: Reaktion, 2016).

38. See Shannon Mattern, *Deep Mapping the Media City* (Minneapolis: University of Minnesota Press, 2018), 13–17; and works discussed there including Siegfried Zielinski, *Deep Time of the Media: Toward an Archaeology of Hearing and Seeing by Technical Means* (Cambridge, Mass.: MIT Press, 2006); and Christopher L. Witmore, "Symmetrical Archaeology: Excerpts of a Manifesto," *World Archaeology* 39, no. 4 (2007): 546–62.

39. See the introduction to *Old Media and the Medieval Concept: Media Ecologies before Early Modernity,* ed. Thora Brylowe and Stephen Yeager (Montreal: Concordia University Press, 2021); and Ingrid Nelson, "Ambient Media and Chaucer's *House of Fame,*" ELH 88, no. 3 (2021): 551–78.

40. From the foreword to *Prosthesis in Medieval and Early Modern Culture,* ed. Chloe Porter, Katie L. Walter, and Margret Healy (New York: Routledge, 2018).

41. Hankins and Silverman, *Instruments,* 13.

42. Michael D. Gordon, *Scientific Babel: How Science Was Done before and after Global English* (Chicago: University of Chicago Press, 2015).

43. Behind these remarks are important concepts in science studies, including Andrew Pickering's "mangle of practice," Bruno Latour's "actor network theory," Karen Barad's "agential realism," and especially Isabelle Stengers's "cosmopolitics"—each indicating how knowledge is mediated by many persons and things, caught up in the vast world. See, besides others, Andrew Pickering, *The Mangle of Practice: Time, Agency, and Science* (Chicago: University of Chicago Press, 1995); Bruno Latour, *Reassembling the Social: An Introduction to Actor-Network-Theory* (Oxford: Oxford University Press, 2005); and Isabelle Stengers, *Cosmopolitics I,* trans. Robert Bononno (Minneapolis: University of Minnesota Press, 2010).

44. See Karla Mallette, *Lives of the Great Languages: Arabic and Latin in the Medieval Mediterranean* (Chicago: University of Chicago Press, 2021); Christian Etheridge and Michele Campopiano, eds., *Medieval Science in the North: Travelling Wisdom, 1000–1500* (Turnhout, Belgium: Brepols, 2021).

ONE. INTELLIGENT OBJECTS

1. Geoffrey Chaucer, *A Treatise on the Astrolabe,* ed. Sigmund Eisner, in *The Prose Treatises,* Variorum Edition of the Works of Geoffrey Chaucer 6/1 (Norman: University of Oklahoma Press, 2002). All citations to the treatise will appear in parentheses.

2. For evidence of fifteenth-century readership, see Edgar Laird, "Chaucer and Friends: The Audience for *The Treatise on the Astrolabe,*" *The Chaucer Review* 41, no. 4 (2007): 439–44; Simon Horobin, "The Scribe of Bodleian Library MS Bodley 619 and the Circulation of Chaucer's *Treatise on the Astrolabe,*" *Studies in the Age of Chaucer* 31 (2009): 109–24. Like the contemporary *Equatorie of the Planetis,* Chaucer's *Astrolabe* would satisfy a growing coterie whose appetite for applied science was apparent. An English instruction manual on how to make and use another astronomical device, *The Equatorie of the Planetis* was once attributed to Chaucer. For discussion, see Kari Anne Rand, "The Authorship of *The Equatorie of the Planetis* Revisited," *Studia Neophilologica* 87 (2015): 15–35.

3. Horobin, "Scribe," 109.

4. Edward Grant, *The Nature of Natural Philosophy in the Late Middle Ages* (Washington, D.C.: Catholic University of America Press, 2010), 163–94. Jonathan Sawday argues that early technoculture involves a "work of imagination" (*Engines of the Imagination: Renaissance Culture and the Rise of the Machine* [London: Routledge, 2007], xviii).

5. Roger Bacon, *Fr. Rogeri Bacon Opera quaedam hactenus inedita*: "Experimentator tamen fidelis et magnificus, ad hoc anhelat, ut ex tali materia fieret, et tanto artificio, quod naturaliter coelum motu diurno volveretur" (ed. J. S. Brewer [1859; repr. Cambridge: Cambridge University Press, 2012], 537).

6. Derek de Solla Price, "Philosophical Mechanism and Mechanical Philosophy: Some Notes Towards a Philosophy of Scientific Instruments," *Annali dell' Instituto et Museo di Storia di Scienza di Firenze* 5, no. 1 (1980): 76.

7. See the useful remarks on *poiesis* in E. R. Curtius, *European Literature and the Latin Middle Ages,* trans. Willard Task (London: Routledge, 1953), 145–47. A sharp treatment of poetic figures and forms can be found in Lisa Cooper, "Figures for 'Gretter Knowing': Forms in the *Treatise on the Astrolabe,*" in *Chaucer and the Subversion of Form,* ed. Thomas A. Prendergast and Jessica Rosenfeld (Cambridge: Cambridge University Press,

2018), 99–124, which should complement my account of the eloquence of the technical ensemble.

8. Elizabeth Hamm, "Modeling the Heavens: *Sphairopoiia* and Ptolemy's *Planetary Hypotheses*," *Perspectives on Science* 24, no. 4 (2016): 416–24.

9. Stephen G. Nichols, "The New Medievalism: Tradition and Discontinuity in Medieval Culture," in *The New Medievalism,* ed. Marina S. Brownlee, Kevin Brownlee, and Stephen G. Nichols (Baltimore, Md.: Johns Hopkins University Press, 1991), 4.

10. Arianna Borrelli, *Aspects of the Astrolabe: "Architectonica Ratio" in Tenth- and Eleventh-Century Europe* (Stuttgart: Franz Steiner Verlag, 2008), 17–18. Elsewhere, Borrelli associates the astrolabe more broadly with the development of a "flat sphere," which is both "an instrument for the hands, like a pair of compasses, and one for the mind, like the rules of logic" ("The Flat Sphere," in *Variantology: On Deep Time Relations of Arts, Sciences, and Technologies,* ed. Siegfried Zielinski and David Link [Cologne: Walther König, 2006], 146). I also benefit from databases and catalogue descriptions of surviving instruments, including Epact: Scientific Instruments of Medieval and Renaissance Europe, under the direction of Jim Bennett, www.mhs.ox.ac.uk/epact/, accessed March 24, 2018; F. A. B. Ward, *A Catalogue of European Scientific Instruments in the Department of Medieval and Later Antiquities of the British Museum* (London: British Museum, 1981); Koenraad van Cleempoel and Silke Ackermann, eds., *Astrolabes at Greenwich: A Catalogue of the Astrolabes in the National Maritime Museum, Greenwich* (Oxford: Oxford University Press, 2005); Roderick Webster and Marjorie Webster, *Western Astrolabes: Historic Scientific Instruments of the Adler Planetarium & Astronomy Museum,* vol. 1 (Chicago: Adler Planetarium, 1998); Robert William T. Gunther, *Astrolabes of the World,* 2 vols. (Oxford: Oxford University Press, 1932); Sharon Gibbs with George Saliba, *Planispheric Astrolabes from the National Museum of American History* (Washington, D.C.: Smithsonian, 1984); David King, *Astrolabes from Medieval Europe,* Variorum Collected Studies (New York: Routledge, 2011), part 12.

11. Stephen McCluskey, *Astronomies and Cultures in Early Medieval Europe* (Cambridge: Cambridge University Press, 1998). See also Borrelli, *Aspects of the Astrolabe,* 63–79.

12. Charles Burnett, "*Algorismi vel helcep decentior est diligentia:* The Arithmetic of Adelard of Bath and His Circle," in *Mathematische Probleme im*

Mittelalter: der lateinische und arabische Sprachbereich, ed. Menso Folkerts (Wiesbaden: Harrassowitz, 1996); repr. in *Numerals and Arithmetic in the Middle Ages* (Burlington, Vt.: Ashgate, 2010), 221–31.

13. See Bernard Hamilton, "Knowing the Enemy: Western Understanding of Islam at the Time of the Crusades," *Journal of the Royal Asiatic Society,* 3rd ser., 7, no. 3 (1997): 373–87.

14. For details about the development of an astronomy curriculum and defined corpus see Olaf Pedersen, "The Corpus Astronomicum and the Traditions of Medieval Latin Astronomy," in *Colloquia Copernicana III,* ed. Owen Gingerich and Jerzy Dobrzycki (Wrocław: Ossolineum, 1975), 57–96; Pedersen, "The Origins of the 'Theorica Planetarum,'" *Journal for the History of Astronomy* 12 (1981), 113–23; Lynn Thorndike, *The "Sphere" of Sacrobosco and Its Commentators* (Chicago: University of Chicago Press, 1949), 42–43. For a description of the contents of manuscripts, see Kari Anne Schmidt, *The Authorship of "The Equatorie of the Planetis,"* 103–49 and 186–282; and Linne R. Mooney, *The Index of Middle English Prose: Handlist XI, Manuscripts in the Library of Trinity College, Cambridge* (Woodbridge, UK: D. S. Brewer, 1995), 109–21. For a note on the *bismillah* translation, see D. J. Price, ed., *The Equatorie of the Planetis: Edited from Peterhouse MS 75.I* (Cambridge: Cambridge University Press, 1955), 62.

15. Thomas Usk, *The Testament of Love,* ed. R. Allen Shoaf (Kalamazoo, Mich.: Medieval Institute Publications, 1998), prol. 23–27; for Giles of Rome, see T. Matthew N. McCabe, *Gower's Vulgar Tongue: Ovid, Lay Religion, and English Poetry in the "Confessio Amantis"* (Cambridge: D. S. Brewer, 2011), 77.

16. As Seb Falk shows, an English work on the equatorium composed by John Westwyk—who may have been influenced by Chaucer's *Astrolabe*—introduces "several words or phrases that may appear in English for the first time," including *diametral, eccentrik, geometrical,* and *precise* ("Vernacular Craft and Science in the *Equatorie of the Planetis,*" *Medium Aevum* 88, no. 2 [2019]: 344–45). Westwyk also employs *almenak, aryn,* and *alhudda,* words of Arabic origin.

17. For example, *De Algorismo* in Trinity College Dublin, MS 367, fol. 64. See also Charles Burnett, "Manuscripts of Latin Translations of Scientific Texts from Arabic," in *Taxonomies of Knowledge: Information and Order in Medieval Manuscripts,* ed. Emily Steiner and Lynn Ransom (Philadelphia: Schoenberg Institute, 2015), 80–89; Cornelius O'Boyle, "Astrology

and Medicine in Later Medieval England: The Calendars of John Somer and Nicholas of Lynn," *Sudhoffs Archiv* 89, no. 1 (2005), 3–5; John Somer, *The Kalendarium of John Somer,* ed. Linne R. Mooney (Athens: University of Georgia Press, 1999), 35; Paul Acker, "*The Craft of Nombrynge* in Columbia University Library, Plimpton MS 259," *Manuscripta* 37 (1993): 71–83; Robert Steele, ed. *The Earliest Arithmetics in English,* EETS e.s. 118 (London: H. Milford, 1922).

18. Jenna Mead, "Geoffrey Chaucer's *Treatise on the Astrolabe,*" *Literature Compass* 3, no. 5 (2006): 985. Elsewhere, Mead wisely observes that the traces of a "textual genealogy" should prevent our taking the work as "an unmarked, white text" ("Reading by Said's Lantern: Orientalism and Chaucer's *Treatise on the Astrolabe,*" *Al-Masaq* 15, no. 1 [2003]: 79).

19. See Tim William Machan, "The Ecology of Middle English," in *English in the Middle Ages* (Oxford: Oxford University Press, 2005), 1–20.

20. See Jonathan Hsy, *Trading Tongues: Merchants, Multilingualism, and Medieval Literature* (Columbus: Ohio State University Press, 2013), 57.

21. See Michael D. Gordon, *Scientific Babel: How Science Was Done before and after Global English* (Chicago: University of Chicago Press, 2015).

22. See Karla Mallette, *Lives of the Great Languages: Arabic and Latin in the Medieval Mediterranean* (Chicago: University of Chicago Press, 2021).

23. See Sharon Kinoshita's "Deprovincializing the Middle Ages," in *The Worlding Project: Doing Culture Studies in the Era of Globalization,* ed. Rob Wilson and Christopher Leigh Connery (Santa Cruz, Calif.: New Pacific, 2007), 61–75; John Dagenais and Margaret R. Greer, eds., "Decolonizing the Middle Ages," special issue, *Journal of Medieval and Early Modern Studies* 30, no. 3 (2000); and the Global Middle Ages Project (GMAP), founded by Susan Noakes and Geraldine Heng, globalmiddleages.org. Anticolonialist science education is ongoing: Chanda Prescod-Weinstein, "Decolonising Science Reading List," medium.com/@chanda/decolonising-science-reading-list-339fb773d51f, accessed April 20, 2017. For more on medieval cosmopolitanism see the essays in John M. Ganim and Shayne Aaron Legassie, eds., *Cosmopolitanism in the Middle Ages* (New York: Palgrave Macmillan, 2013); and in Kathryn L. Lynch, ed., *Chaucer Cultural Geography* (New York: Routledge, 2002); and see the seminal essay of Elizabeth Salter, "Chaucer and Internationalism," *Studies in the Age of Chaucer* 2 (1980): 71–79. After I had published an earlier version of the chapter, there appeared a rich new essay by

Christine Chism, "Transmitting the Astrolabe: Chaucer, Islamic Astronomy, and the Astrolabic Text," in *Medieval Textual Cultures: Agents of Transmission, Translation and Transformation,* ed. Faith Wallis and Robert Wisnovsky (Berlin: De Gruyter, 2016), 85–120, working toward similar conclusions about the ecumenical and cross-cultural astrolabe text.

24. The following analysis of the astrolabic science as an imaginative projection—something of a realized ideal, or concrete abstraction—should serve to complement Daniel Wakelin's *Immaterial Texts in Late Medieval England: Making English Literary Manuscripts, 1400–1500* (Cambridge: Cambridge University Press, 2022). Although dealing with the construction and transmission of imaginative literary texts, Wakelin offers a relevant and sophisticated account of nontangible systems, conventions, attitudes, and ideas, including geometrical ones, that go into the making and marking of the manuscript page.

25. Borrelli, *Aspects of the Astrolabe,* 47.

26. Those mirroring effects are treasured attributes in later medieval art and architecture (e.g., in symmetrical wings of buildings, rosettes in church windows, and diapering patterns in painting), as discussed in the third chapter of Felipe Cucker, *Manifold Mirrors: The Crossing Paths of the Arts and Mathematics* (Cambridge: Cambridge University Press, 2013).

27. See Glen Van Brummelen, *The Mathematics of the Heavens and the Earth: The Early History of Trigonometry* (Princeton, N.J.: Princeton University Press, 2009), 1–8.

28. See Lynn Thorndike, *A History of Magic and Experimental Sciences,* 8 vols. (New York: Columbia University Press, 1958), 3:405; and his edition of *The "Sphere" of Sacrobosco and Its Commentators* (Chicago: University of Chicago Press, 1949), 78, 119; and Borrelli, *Aspects of the Astrolabe,* 160–61.

29. See Somer, *Kalendarium,* 3. For an overview of the matter, see José Chabás, "Characteristics and Typologies of Medieval Astronomical Tables," *Journal of the History of Astronomy* 43 (2012): 269–86.

30. See Epact, 40428; Ward, *Catalogue,* 110–13.

31. See Cooper, "Figures for 'Gretter Knowing,'" 114–17, where she observes the emphasis on "equivalence" and "universality" available to practitioners. For Cooper, the effect of resisting hierarchies is a temporary effect and ultimately revolves around Chaucer.

32. Bruno Latour, *An Inquiry into Modes of Existence: An Anthropology of the Moderns,* trans. Catherine Porter (Cambridge, Mass.: Harvard University

Press, 2013), 76. On flat ontology and topology, see also Latour, *Reassembling the Social: An Introduction to Actor-Network-Theory* (New York: Oxford University Press, 2005); Manuel De Landa, *Intensive Science and Virtual Philosophy* (London: Bloomsbury, 2002), 51.

33. Compare Catherine Eagleton, *Monks, Manuscripts, and Sundials: The Navicula in Medieval England* (Leiden: Brill, 2010), 86: "The navicula . . . could remind its owners of the good, Christian, life, and their passage through life on the way to the New Jerusalem. Running alongside the religious interests of late medieval England were its military and trading interests, and here, too, we can see that the navicula could have had a strong iconographic significance for its owners. England had become rich thanks to the wool trade, and this new wealth enriched many towns and ports in England." In two of the earliest manuscripts (MS Bodley 68 and Trinity College O.5.26), the device is called a "Little Ship of Venice," *navicula de venetiis* or "lytel schippe of venyse," which is to reference a great international shipping power and the sort of commercial interchange that that connection probably implies, as suggested by, among others, Derek J. Price, "The Little Ship of Venice—A Middle English Instrument Tract," *Journal of the History of Medicine and Allied Sciences* 15 (1960): 399–407. More specifically, the allusion may attest to the technological, commercial, and cultural achievements contemporaries increasingly came to associate with the northern Italian trading ports, and with the ships they increasingly used to carry not only grain, salt, fish, and hides but also enslaved people packed away in the deep holds of Venetian ships—not just any ships, but specifically cogs or single-masted roundships resembling the shape of the navicula. The later medieval cog is not a mere emblem of local seafaring, but a brute fact in the ongoing and emerging history of slave economy in the Mediterranean and the Baltic and Black Seas, as if reckoning with the times. My thanks to Kate Baxter for exploring the idea of the instrument passing through time-space, and to Colin Keohane for his research assistance on contemporary ship types and technology.

34. See Cooper, "Figures for 'Gretter Knowing,'" 108.

35. Owen Gingerich, "Zoomorphic Astrolabes and the Introduction of Arabic Star Names into Europe," *Annals of the New York Academy of Sciences* 500, no. 1 (1987): 95.

36. *Middle English Dictionary*, s.v. *calle*.

37. Talk of promiscuous and monogamous networks, male and female connectors, and associated technological capacities as a "*matrix,* a womb, the mother-matter that spawns us all" persist in computational vocabularies; see Erik Davis, *TechGnosis: Myth, Magic, and Mysticism in the Age of Information* (New York: Harmony, 1998), 326; Wendy Hui Kyong Chun, "Habitual New Media: Exposing Empowerment," Barnard College, New York City, October 10, 2013, vimeo.com/78287998, accessed August 29, 2016.

38. BL, MS Arundel 377. See Charles Homer Haskins, "Adelard of Bath and Henry Plantagenet," *English Historical Review* 28 (1913): 515–16; Charles Burnett, *The Introduction of Arabic Learning into England* (London: British Library, 1997), 31–69.

39. Jane Bennett, *Vibrant Matter: A Political Ecology of Things* (Durham, N.C.: Duke University Press, 2010), 98–99. For an able defense of the critical possibilities of anthropomorphism, see Tom Tyler, *Ciferae: A Bestiary in Five Fingers* (Minneapolis: University of Minnesota Press, 2012), 51–64.

40. Seth Lerer argues that Chaucer's *Astrolabe* exhibits a conspicuous genealogical cast ("Chaucer's Sons," *University of Toronto Quarterly* 73, no. 3 [2004]: 909). Joe Stadolnik has argued that "Bread and Milk for Children" should be taken less literally in manuscript contexts to apply to the reading diet of Latinate men, not only (or even) children ("Little Lewis and Latin Folk in Chaucer's Prologue to the *Treatise on the Astrolabe*," *Journal of English and Germanic Philology* 121, no. 3 [2022]: 359–82). All to say, Chaucer's work can be construed as part of a tradition of intellectual and national paternalism that occludes female involvement, where the labor of paternal authorship generates a seminal work in translation, perpetuating a presumptive heteronormative order.

41. Mead, "Geoffrey Chaucer's *Treatise on the Astrolabe*," 980.

42. See, for example, Albertus Magnus, *Book of Minerals,* trans. Dorothy Wyckoff (Oxford: Clarendon Press, 1967), 169, 196. For an astute analysis of rock as the matrix of metal in Albertus, see Valerie Allen, "Mineral Virtue," in *Animal, Vegetable, Mineral: Ethics and Objects,* ed. Jeffrey Jerome Cohen (Washington, D.C.: Oliphaunt, 2012), 133–34. Her explanation of how metaphors matter in scientific discourse is apt: "Just as there can come a point when a scientific model so aptly illustrates its phenomena that it steps out of the realm of metaphor to describe rather than

fictionalize those phenomena, so the reproductive model stops analogizing and starts describing" (140).

43. Manuel De Landa, *A Thousand Years of Nonlinear History* (Cambridge, Mass.: MIT Press, 1997); Jussi Parikka, *A Geology of Media* (Minneapolis: University of Minnesota, 2015), 4. Such elemental vitality is akin to Gilles Deleuze and Félix Guattari's "machinic phylum" (*A Thousand Plateaus,* vol. 2 of *Capitalism and Schizophrenia,* trans. Brian Massumi [Minneapolis: University of Minnesota Press, 1980], 409). See also De Landa, "Nonorganic Life," in *Incorporations,* ed. Jonathan Crary and Sanford Kwinter (New York: Zone, 1992), 135–36; De Landa, "The Machinic Phylum," in *TechnoMorphica,* ed. Joke Brouwer and Carla Hoekendijk (Rotterdam: V2, 1997), consulted in online version without pagination at v2.nl/archive/articles/the-machinic-phylum, accessed August 1, 2016.

44. See Patricia Clare Ingham, *The Medieval New: Ambivalence in an Age of Innovation* (Philadelphia: University of Pennsylvania Press, 2015), 86.

45. Gilbert Simondon, *On the Mode of Existence of Technical Objects* (Minneapolis: University of Minnesota Press, 2017), 72–74, 58, 252–53.

46. Nathan Sidoli and J. L. Berggren, "The Arabic version of Ptolemy's *Planisphere* or *Flattening the Surface of the Sphere:* Text, Translation, Commentary," SCIAMVS 8 (2007): 126.

47. See Elly Dekker, *Illustrating the Phaenomena: Celestial Cartography in Antiquity and the Middle Ages* (Oxford: Oxford University Press, 2012), 27; Jérôme Bonnin, "Conarachne et Pelecinum: About Some Graeco-Roman Sundial Types," *British Sundial Society Bulletin* 27, no. 2 (2015): 28–32. And, for a possible reconstruction of the *arachne,* see Erkka Maula, "The Spider in the Sphere: Eudoxus' 'Arachne,'" *Philosophia* 5–6 (1975–76): 225–57.

48. J. D. North, *Cosmos: An Illustrated History of Astronomy and Cosmology* (Chicago: University of Chicago Press, 2008), 192.

49. See, for instance, evidence discussed in E. Ruth Harvey, "The Swallow's Nest and Spider's Web," in *Studies in English Language and Literature: "Doubt Wisely" (Papers in Honour of E. G. Stanley)* (London: Routledge, 1996), 332–33.

50. Albertus Magnus, *De animalibus libri XXVI,* ed. Hermann Stadler, 2 vols. (Munster: Aschendorffsche, 1916–20), 1:629 (bk. 8, tract. 4, ch. 1), 2:1582 (bk. 26); in *On Animals: A Medieval Summa Zoologica,* trans. Kenneth

F. Kitchell Jr. and Irven Michael Resnick, 2 vols. (Baltimore, Md.: Johns Hopkins University Press, 1999), 1:729, 2:1744.

51. Bartholomaeus Anglicus, *On the Properties of Things: John Trevisa's Translation of "De proprietatibus rerum of Bartholomaeus Anglicus": A Critical Text,* vol. 2, ed. M. C. Seymour (Oxford: Clarendon, 1975–88), 1139 (18.11).

52. Bartholomaeus Anglicus, *On the Properties of Things,* 1139 (18.11).

53. Pliny, *Natural History,* vol. 3, *Books 8–11,* trans. H. Rackham, Loeb Classical Library 353 (Cambridge, Mass.: Harvard University Press, 1940), 481–83 (11.28).

54. Katarzyna Michalski and Sergiusz Michalski, *Spider* (London: Reaktion, 2010), 58–64.

55. Harvey suggests: "When admirers of the cobweb start to talk about geometry they are getting dangerously near the liberal arts, those skills proper to man" ("Swallow's Nest and Spider's Web," 334).

56. See Albertus Magnus, *On Animals,* 1:486 (bk. 4, tract. 4), 2:729 (bk. 8, tract. 4), 2:1744 (bk. 26) (*De animalibus,* 1:405 [bk. 4, tract. 2, ch. 4], 1:629 [bk. 8, tract. 4, ch. 1], 2:1582–86 [bk. 26]).

57. On her *instrumenta venandi,* or hunting implements, see Albertus Magnus, *On Animals,* 2:1740 (bk. 21, tract. 8; *De animalibus,* 2:1579 [bk. 26]).

58. Jakob von Uexküll, "The Theory of Meaning," *Semiotica* 42, no. 1 (1982): 42, 66.

59. Tim Ingold, *Being Alive: Essays on Movement, Knowledge and Description* (New York: Routledge, 2011), 91. Ingold's SPIDER does not embody quite the same methodology as Latour's ANT (Actor Network Theory). As SPIDER says to ANT in a playful colloquy, the web is dynamic: "They are the lines *along* which I live, and conduct my perception and action in the world. For example, I know when a fly has landed in the web because I can feel the vibrations in the lines through my spindly legs, and it is along these same lines that I run to retrieve it."

60. See Parikka's account of "insect media," which shows how ultramodern technoculture is on a continuum with a "technics of nature." Accordingly, "the issue of categorical difference between animals and humans, nature and technology is bracketed and the view of affects, movements, and relations among parts is posited as primary" (*Insect Media: An Archaeology of Animals and Technology* [Minneapolis: University of Minnesota Press, 2010], 72).

61. Andy Clark, *Natural-Born Cyborgs: Minds, Technologies, and the Future of Human Intelligence* (Don Mills, Ont.: Oxford University Press, 2004), 11 and passim; Compare the notion of the "alphabetic body" in Brian Rotman, *Becoming beside Ourselves: The Alphabet, Ghosts, and Distributed Human Being* (Durham, N.C.: Duke University Press, 2008), 13–31.

62. Roger Bacon, *De Scientia Perspectiva*: "Nam altitudo earum super horizonta, et magnitudo, et figura, et multitudo, et omnia quae in eis sunt, certificantur per modos videndi in instrumentis" (*The "Opus Majus" of Roger Bacon*, vol. 2, ed. John Henry Bridges [Cambridge: Cambridge University Press, 1897; repr. 2010], 2).

TWO. GRAPHIC INTERFACES

1. All subsequent citations of Chaucer's fictional works are drawn from *The Riverside Chaucer*, ed. Larry Benson, 3rd ed. (Boston: Houghton Mifflin, 1987), and appear in parentheses.

2. Marijane Osborne, *Time and the Astrolabe in the "Canterbury Tales"* (Norman: University of Oklahoma Press, 2002), 49–50. Chaucer could have used the instrument to check whether the constellations were arranged so that the sun rose over the horizon ahead of the "Eagle star Altair."

3. As I have detailed elsewhere, the poetic vision takes after a two-dimensional graphed astrolabe plate. Chaucer's eagle flies within the neighborhood of two stellar "eagles," which locates the poet between the branches of a standard *Y*-shaped rete on contemporary astrolabes, on left side of which perches Altair (Alpha Aquilae, or Arabic *al-nasr al-tair* for "flying eagle") and, on the right, "Wega" (Alpha Lyrae, Latinizing *al-nasr al-wāqi*, "swooping eagle"). In diagrams of the rete in manuscript copies of Chaucer's *Astrolabe*, both fixed stars are sometimes labeled and, in the case of Oxford, Bodleian Library, MS e. Museo 54, fol. 2v, and Cambridge University Library, MS Dd.III.53, fol. 213v, Vega is figured as a bird.

4. Adelard of Bath, *Libellus de opere astrolapsus,* cited from Bruce G. Dickey, "Adelard of Bath: An Examination Based on Heretofore Unexamined Manuscripts" (PhD diss., University of Toronto, 1982), 172. See also Barbara Obrist, "The Idea of a Spherical Universe and Its Visualization in the Earlier Middle Ages (Seventh to Twelfth Century)," in *The Visualization of Knowledge in Medieval and Early Modern Europe,* ed. Marcia Kupfer et al. (Turnhout, Belgium: Brepols, 2020), 250.

5. Astrolabe volvelles are found, for instance, in Oxford, Bodleian Library, MS Ashmole 391 and MS Ashmole 369; and in London, British Library, MS Egerton 848. See further Laurel Braswell-Means, "The Vulnerabilty of Volvelles in Manuscript Codices," *Manuscripta* 35 (1991): 43–54; Gianfranco Crupi, "Volvelles of Knowledge: Origin and Development of an Instrument of Scientific Imagination (13th–17th centuries)," *JLIS.it* 10, no. 2 (2019): 1–27.

6. Catherine Eageleton, "John Wethamstede, Abbot of St. Albans, on the Discovery of the Liberal Arts and Their Tools: Or, Why Were Astronomical Instruments in Late-Medieval Libraries?" *Mediaevalia* 29, no. 1 (2008): 119.

7. Here I am thinking of the instruction manual in relation to the document theory of Maurizio Ferraris, *Documentality: Why It Is Necessary to Leave Traces,* trans. Richard Davies (New York: Fordham University Press, 2012), 202. On the genealogy of document theory, see Niels Windfeld Lund, "Document Theory," *Annual Review of Information Science and Technology* 43 (2009): 399–432.

8. For an extended treatment of the "graphic field" of manuscripts, see Mary C. Olson, *Fair and Varied Forms: Visual Textuality in Medieval Illustrated Manuscripts* (New York: Routledge, 2003).

9. Elaine Treharne, *Perceptions of Medieval Manuscripts: The Phenomenal Book* (Oxford: Oxford University Press, 2021), 1. Various schemes are possible. Among the main features worth mentioning is the way written content is ordered, divided, glossed, and cross-referenced to become usable, forming the instrumental *ordinatio* and *mise-en-page*. See further M. B. Parkes, "The Influence of the Concepts of *Ordinatio* and *Compilatio* on the Development of the Book," in *Scribes, Scripts, and Readers* (London: Hambledon, 1991), 35–70.

10. See, for example, the ninth-century copy divided in two volumes in British Library, Harley MS 3941 and 3941/2. For more medieval and early modern examples, see Mary A. Rouse and Richard H. Rouse, "The Development of Research Tools in the Thirteenth Century," in *Authentic Witnesses: Approaches to Medieval Texts and Manuscripts* (Notre Dame, Ind.: University of Notre Dame Press, 1991), 221–55; Ann M. Blair, *Too Much to Know: Managing Scholarly Information before the Modern Age* (New Haven, Conn.: Yale University Press, 2010); David McKitterick, *Print, Manuscript, and the Search for Order, 1450–1830* (Cambridge:

Cambridge University Press, 2003); Siân Echard, "Pre-Texts: Tables of Contents and the Reading of John Gower's *Confessio Amantis*," *Medium Aevum* 66 (1997): 270–87.

11. On the evolution of the table of contents, see the "Analytic Table of Contents" in *The Etymologies of Isidore of Seville*, ed. and trans. Stephen A. Barney, W. J. Lewis, J. A. Beach, and Oliver Berghof (Cambridge: Cambridge University Press, 2006), 34.

12. Emily Steiner, *John Trevisa's Information Age: Knowledge and the Pursuit of Literature, c. 1400* (Oxford: Oxford University Press, 2021), 106.

13. See Ayelet Even-Ezra, *Lines of Thought: Branching Diagrams and the Medieval Mind* (Chicago: University of Chicago Press, 2021), 89–118 and 156–62.

14. Peterhouse, Cambridge, MS 75.1, f. 71v, as transcribed in *The Equatorie of the Planetis: Edited from Peterhouse MS 75.1*, ed. D. J. Price (Cambridge: Cambridge University Press, 1955), 18. For an astute analysis of "instrument craft," see Seb Falk, "Vernacular Craft and Science in the *Equatorie of the Planetis*," *Medium Aevum* 88, no. 2 (2019): 329–60.

15. MS Aberdeen University 123, fol. 44v, as transcribed in appendix 7 of Catherine Eagleton, *Monks, Manuscripts, and Sundials: The Navicula in Medieval England* (Leiden: Brill, 2010), 247.

16. A Middle English treatise on the navicula in Cambridge, Trinity College, MS O.5.26, 115r, refers to geometrical templates as "instrumentes," and moreover, standalone drawings of the device are labeled *instrumentum* in Cambridge University Library, MS Ee.III.61, fols. 191–93. On the distinction between instrument and figure in these cases, see Eagleton, *Monks, Manuscripts, and Sundials*, 61–65.

17. Some time ago, Jill H. Larkin and Herbert A. Simon explored the differences between lingual and diagrammatic descriptions of a hypothetical rope-and-pulley system: while verbal description and visual diagram might be roughly equivalent informationally, the diagram itself seems to format data to be the more efficiently searched, recognized, and activated ("Why a Diagram Is [Sometimes] Worth Ten Thousand Words," *Cognitive Science* 11 [1987]: 65–99).

18. Valeria Giardino, "Behind the Diagrams: Cognitive Issues and Open Problems," in *Thinking with Diagrams: The Semiotic Basis of Human Cognition*, ed. Sybille Krämer and Christina Ljungberg (Boston: De Gruyter, 2016), 81.

19. Sybille Krämer and Christina Ljungberg, "Thinking and Diagrams—An Introduction," in *Thinking with Diagrams,* 1.

20. Jeffrey F. Hamburger, "Mindmapping: The Diagram as Paradigm in Medieval Art—and Beyond," in Kupfer, *Visualization of Knowledge,* 61–86; Hamburger, *Diagramming Devotion* (Chicago: University of Chicago Press, 2019).

21. Even-Ezra, *Lines of Thought,* 50–83.

22. Even-Ezra found that fifty-two out of a sample of seventy-one manuscripts contain horizontal tree diagrams, suggesting a common visualization strategy among scribes and readers of the *Organon* (*Lines of Thought,* 50–83).

23. Krämer and Ljungberg, "Thinking and Diagrams," 2–4.

24. Bogen, "The Diagram as Board Game," in Krämer and Ljungberg, *Thinking with Diagrams,* 189.

25. Bogen, 195.

26. Bogen, 206. Apropos, Johan Huizinga famously contends of *homo ludens* that "culture arises in the form of play. . . . [and] is played from the very beginning" (*Homo Ludens: A Study of the Play Element in Culture* [London: Routledge & Kegan Paul, 1950], 46).

27. Krämer and Ljungberg, "Thinking and Diagrams," 12.

28. Peter Barber and Catherine Delano-Smith, "Maps in Medieval Europe," in *The Routledge Handbook of Mapping and Cartography,* ed. Alexander Kent and Peter Vujakovic (New York: Routledge, 2018), 119.

29. Maximos Planudes as cited in Marcia Kupfer, "The Rhetoric of World Maps in Late Antiquity and the Middle Ages," in *Visualization of Knowledge,* 261.

30. Kupfer, 263.

31. Matthew Boyd Goldie, *Scribes of Space: Place in Middle English Literature and Late Medieval Science* (Ithaca, N.Y.: Cornell University Press, 2019). In his examples, Canterbury Cathedral is characterized by buildings, walls, gates, ditches, and plumbing systems, abstracting individual elements from the general setting in a way that is "unrealistic and nonrepresentational," but nevertheless an index to a physical domain (35). Sherwood Forest is partly delineated by means of quadrants and triangles, columns and rows, dividing and distributing a well-known woodland area in a reliable manner that can orient observers on the ground (40–42).

32. Daniel K. Connolly, "Pilgrimage in the Itinerary Map of Matthew Paris," *The Art Bulletin* 81, no. 4 (1999): 598–622.

33. Hamburger, *Diagramming Devotion,* 248–50.

34. Hamburger, 257–59. Felipe Cucker demonstrates that a diagrammatic imagination informed much church art and architecture, forming treasured attributes of the built environment (e.g., the symmetrical wings of cathedrals, rosettes in church windows, and diapering patterns in wall painting), so that they may be said to equip liturgy and other forms of priestcraft, as explored in the third chapter of Felipe Cucker, *Manifold Mirrors: The Crossing Paths of the Arts and Mathematics* (Cambridge: Cambridge University Press, 2013).

35. Valerie Allen, "To Measure Is to Feel: The Mathematics of Middle English Metric Relics," *Journal of Medieval and Early Modern Studies* 52, no. 2 (2022): 234–36.

36. Hamburger, "Mindmapping," 74.

37. Catarina Dutilh Novaes, *Formal Languages in Logic: A Philosophical and Cognitive Analysis* (Cambridge: Cambridge University Press, 2012), 79; Novaes cites Clark N. Glymour, *Thinking Things Through: An Introduction to Philosophical Issues and Achievements* (Cambridge, Mass.: MIT Press, 1997), 71.

38. Even-Ezra, *Lines of Thought,* 47, quoting from a passage of Thomas Le Myésier's *Breviculum,* which begins: "Ratio, quare has figuras visibiles facio, stat in hoc, ut, cum videntur statim aperte, multa per visionem earum ad memoriam simul et semel reducuntur."

39. Larkin and Simon, "Why a Diagram," 98.

40. Giardino, "Behind the Diagrams," 98.

41. Crupi, "Volvelles of Knowledge," 2.

42. Crupi, 3–5.

43. Connolly, "Pilgrimage," 610.

44. Goldie, *Scribes of Space,* 94.

45. Brian Rotman, *Mathematics as Sign: Writing, Imagining, Counting* (Stanford, Calif.: Stanford University Press, 2000), 57.

46. Maurice Merleau-Ponty, *Phenomenology of Perception,* trans. Colin Smith (London: Routledge, 1962), 443, cited in Brian Rotman, *Becoming Beside Ourselves: The Alphabet, Ghosts, and Distributed Human Being* (Durham, N.C.: Duke University Press, 2008), 34.

47. Gilles Châtelet, *Figuring Space: Philosophy, Mathematics and Physics*, trans. Robert Shore and Muriel Zagha (Dordrecht: Kluwer Academic, 2000), 10.

48. Rotman, *Mathematics as Sign*, 39. A broader context in which to consider mathematical notation is that of medieval manuscript transmission, the product of many tangible and nontangible systems, conventions, and ideas, as elucidated in Daniel Wakelin's *Immaterial Texts in Late Medieval England: Making English Literary Manuscripts, 1400–1500* (Cambridge: Cambridge University Press, 2022).

49. See further Rotman, *Mathematics as Sign*, 34–35: "Mathematical signs play a *creative* rather than merely descriptive function in mathematical practice . . . so that mathematicians at the same time think their scribbles and scribble their thoughts."

50. Danielle Macbeth, "Seeing How It Goes: Paper-and-Pencil Reasoning in Mathematical Practice," *Philosophia Mathematica* 20 (2012): 60.

51. See Charles Burnett, *Numerals and Arithmetic in the Middle Ages* (Burlington, Vt.: Ashgate, 2010).

52. John North, "Diagram and Thought in Medieval Science," in *Villard's Legacy: Studies in Medieval Technology, Science, and Art in Memory of Jean Gimpel*, ed. Jean Gimple and Marie-Thérèse Zenner (Aldershot, UK: Ashgate, 2004), 265–87.

53. Roger Bacon, *De scientia experimentali*: "Qui vero habet demonstrationem potissimam de triangulo aequilatero sine experientia nunquam adhaerebit animus conclusioni, nec curabit, sed negliget usquequo detur ei experientia per intersectionem duorum circulorum, a quorum alterutra sectione ducantur duae lineae ad extremitates lineae datae; sed tunc recipit homo conclusionem cum omni quiete" (*The "Opus Majus" of Roger Bacon*, vol. 2, ed. John Henry Bridges [1897; repr. Cambridge: Cambridge University Press, 2010], 168; trans. from *The Opus Majus of Roger Bacon*, vol. 2, trans. Robert Belle Burke [1928; repr. New York: Russell & Russell, 1962], 583). For Euclid's verbal and visual demonstrations, see *The First Latin Translation of Euclid's "Elements" Commonly Ascribed to Adelard of Bath*, ed. H. L. L. Busard (Toronto: Pontifical Institute of Mediaeval Studies, 1983), 33–34.

54. Thomas Aquinas, *In duodecim libros Metaphysicorum Aristotelis exposito*, bk. 9, lect. 10, ln. 1888: "Dicit ergo primo, quod diagrammata, idest descriptiones geometriae 'inveniuntur,' idest per inventionem

cognoscuntur secundum dispositionem figurarum in actu. Geometrae enim inveniunt verum quod quaerunt, dividendo lineas et superficies. Divisio autem reducit in actum quod erat in potentia" (ed. M. R. Cathala and Raimondo M. Spiazzi [Turin: Marietti, 1964], 453; partly cited and translated in Hamburger, *Diagramming Devotion,* 83).

55. For context, see G. J. Toomer, "The Chord Table of Hipparchus and the Early History of Greek Trigonometry," *Centaurus* 18, no. 1 (1974): 6–28; Glen Van Brummelen, "Mathematical Tables in Ptolemy's Almagest" (PhD diss., Simon Fraser University, 1993).

56. Hugh of St. Victor, *Practical Geometry,* trans. Frederick A. Homann (Milwaukee, Wisc.: Marquette University Press, 1991), 34; and Roger Baron, ed., *Hugonis de Sancto Victore Opera Propaedeutica* (Notre Dame, Ind.: University of Notre Dame Press, 1966), 17. A brief synopsis of the divisions of geometry is also given in his influential conspectus of the arts: *The Didascalicon of Hugh of St. Victor: A Medieval Guide to the Arts,* trans. Jerome Taylor (New York: Columbia University Press, 1991), 70 (2.10).

57. *Gerberti Opera Mathematica,* ed. Nicolaus Bubnov (Berlin: R. Friedländer & Sons, 1899), 487; noted and translated in Joseph V. Navari, "The Leitmotiv in the Mathematical Thought of Gerbert of Aurillac," *Journal of Medieval History* 1 (1975): 146.

58. "Triangulus . . . planarum principium existit figurarum" (Bubnov, *Gerberti Opera Mathematica,* 71). See the discussion in Nancy Marie Brown, *The Abacus and the Cross: The Story of the Pope Who Brought the Light of Science to the Dark Ages* (New York: Basic), 109.

59. *Hugonis de Sancto Victore Opera Propaedeutica,* 19: "Oportet ergo totius mundi formam huic quodam artificio adaptare . . . ita nunc quoque fieri oportet ut ingentia illa spacia, que humana per se possibilitas comprehendere non sufficit, ratione magistra ad exiguum exemplar deducat, et sic quodammodo sub scientiam coarceat" (modifying the trans. in *Practical Geometry,* 35).

60. Prologue to Richard of Wallingford's *Quadripartitum: Quatuor Tractatus De Corda Versa et Recta et De Sinibus Demonstratis*: "Sinus duplatus est linea recta cuius duo termini sunt simul tantum cum duobus terminis porcionis circuli; et se habent sicut corda et arcus. Sinus rectus est medietas illius corde duplate respectu medietatis illius arcus. Sinus versus est semper illa pars diametri que abscindit tam sinum duplatum quam arcum eius, et est sicut sagitta respectu corde et arcus eius" (*Richard of Wallingford: An*

Edition of His Writings with Introduction, English Translation, and Commentary, vol. 1, ed. and trans. J. D. North [Oxford: Clarendon, 1976], 24–27). Compare *A Source Book in Medieval Science,* ed. Edward Grant (Cambridge, Mass.: Harvard University Press, 1974), 188–89.

61. Richard of Wallingford, *Tractatus rectanguli,* in North, *Richard of Wallingford,* 1:406–31. See description in North's *God's Clockmaker: Richard of Wallingford and the Invention of Time* (New York: Continuum, 2006), 60–61, 349–50.

62. On the "scientific imagination," see Edward Grant, *The Nature of Natural Philosophy in the Late Middle Ages* (Washington, D.C.: Catholic University of America Press, 2010), 163–94; and on how "arguments by analogy were a staple of medieval science and philosophy" as exemplified by the figures of ladder, book, and axe in scientific discourse, see Kellie Robertson, *Nature Speaks: Medieval Literature and Aristotelian Philosophy* (Philadelphia: University of Pennsylvania Press, 2017), 42 and passim.

63. Aristotle, *On Memory* 450a and *On the Soul* 1.1.403a, in *The Complete Works of Aristotle: The Revised Oxford Translation,* ed. Jonathan Barnes, vol. 1 (Princeton, N.J.: Princeton University Press, 1984). For discussion, see Wesley Trimpi, *Muses of One Mind: The Literary Analysis of Experience and Its Continuity* (Eugene, Oreg.: Wipf and Stock, 2009), 380 and passim.

64. So mathematical geometry is arrived at through *abstractio formae,* constructing pure lines, points, and curves that are abstractions from those in matter. See Thomas Aquinas, *Expositio super librum Boethii De trinitate,* quest. 5, art. 3, and quest. 6, art. 2, ed. Bruno Decker (Leiden: Brill, 1955), 186 (ln. 17), 216 (ln. 26); trans. in Thomas Aquinas, *The Division and Methods of the Sciences: Questions V and VI of His Commentary on the "De trinitate" of Boethius,* trans. Armand Maurer, 4th rev. ed. (Toronto: Pontifical Institute of Medieval Studies, 1986), 41 and 78. See also Armand Maurer, "Thomists and Thomas Aquinas on the Foundations of Mathematics," *Review of Metaphysics* 47 (1993): 43–61.

65. Nicole Oresme, *The "Questiones de sphera" of Nicole Oresme,* ed. and trans. Garrett Droppers (PhD diss., University of Wisconsin, Madison, 1966), 14–15, 310–11.

66. *Nicole Oresme and the Medieval Geometry of Qualities and Motions: A Treatise on the Uniformity and Difformity of Intensities Known as "Tractatus de configurationibus qualitatum et motuum,"* ed. and trans. Marshall Clagett (Madison: University of Wisconsin Press, 1968), 164–65: "Etsi nichil sunt

puncta indivisibilia aut linee, tamen oportet ea mathematice fingere pro rerum mensuris et earum proportionibus cognoscendis." Oresme is celebrated for his innovative graphs of intensities in the treatise, where he uses geometrical figures to visualize the relational dynamics of phenomena. On Oresme's contribution to the "evolution of a new image of the world in the period—a world of lines in continual expansion and contraction," see Joel Kaye, *A History of Balance, 1250–1375: The Emergence of a New Model of Equilibrium and Its Impact on Thought* (Cambridge: Cambridge University Press, 2014), 409.

67. John of Salisbury, *The Metalogicon of John of Salisbury: A Twelfth-Century Defense of the Verbal and Logical Arts of the Trivium,* trans. Daniel D. McGarry (Berkeley: University of California Press, 1955), 135; *Ioannis Saresberiensis Metalogicon,* ed. J. B. Hall and K. S. B. Keats-Rohan, Corpus Christianorum Continuatio Mediaevalis 98 (Turnholt, Belgium: Brepols, 1991), 96 (2.20).

68. Bubnov, *Gerberti Opera Mathematica,* 334–35: "Si cujus libet rei altitudinem investigare volueris, hujus modi jaculari ingenio investigare poteris. Sume arcum cum sagitta et filo, et una fili summitate sagittae postremitati inhaerente, altera in manu remanente, sagitta arcu emissa altitudinis mensurandae cacumen percutiat. Post haec alterius fili summatis eodem modo alteri sagittae vel alicui jaculo illigetur, et horum utrumvis projectum altitudinis radicem, ut prius, cacumen feriat. Quo facto, utrumque filum retrahas" (trans. in Hugh, *Practical Geometry,* 79 [appendix c]).

69. Another bow-and-arrow technique is used in the celebrated debate between John Buridan and Nicole Oresme, arguing over the path of an arrow's flight and return to a theoretically rotating earth; see Edward Grant, *Foundations of Modern Science in the Middle Ages: Their Religious, Institutional, and Intellectual Contexts* (Cambridge: Cambridge University Press, 1996), 114–15. From Buridan's perspective, an arrow should not fall to the same place if earth were revolving on an axis in the common understanding. Oresme grants the prima facie difficulty but avers that an arrow shot upwards moves with the rotational direction of the earth and so falls back to the place it left behind. Anticipating a point to which I return, Kaye argues that the method exemplifies a facility for an imagined "escape [from] the gravity of an earth-bound perspective" (*History of Balance,* 461).

70. To diagram the chord is to perceive a latent graphic image much in the way, as Brian Rotman argues, "o" has become both ideogram and

indexical sign for "nothing" (*Mathematics as Sign*, 60). Maurizio Ferraris makes a similar point about units of measurement: "They are as concrete as the *calculi*, the little stones—cones, spheres, marbles, disks, rods, tetrahedral, cylinders—used for calculation" (*Documentality: Why It Is Necessary to Leave Traces*, trans. Richard Davies [New York: Fordham University Press, 2012], 202). Both Rotman and Ferraris insist in different ways that concrete figures co-constitute scientific thinking. Exactly so, mental processes and material traces co-constitute the measures and mathematical models involved in the chord.

71. Rotman, *Mathematics as Sign*, 44.

72. On the variety of images and metaphors that ground mathematical thought, see George Lakoff and Rafael E. Núñez, *Where Mathematics Comes From: How the Embodied Mind Brings Mathematics into Being* (New York: Basic, 2000).

73. See Krämer and Ljungberg, "Thinking and Diagrams," 3–12.

74. North, *Richard of Wallingford*, 1:431.

75. *The "Sphere" of Sacrobosco and Its Commentators*, ed. Lynne Thorndike (Chicago: University of Chicago Press, 1949), 91, 126.

76. Kupfer, "Rhetoric of World Maps," 259–90.

77. Yale University, Beinecke Rare Book and Manuscript Library, MS 404, fol. 6v. This image may be viewed in digital facsimile at Yale University Library, collections.library.yale.edu/catalog/2002755.

78. The Creator as Geometer, in *L'image du monde* in British Library, MS Harley 334, fol. 33v, which may be viewed in digital facsimile at the British Library, bl.uk/manuscripts/Viewer.aspx?ref=harley_ms_334_fs001r, and in *Bible moralisée* in Vienna, Österreichische Nationalbibliothek, Codex Vindobonensis 2554, fol. 1v, which may be viewed in Gerald B. Guest, *Bible moralisée: Codex Vindobonensis 2554, Vienna, Österreichischen Nationalbibliothek* (London: Harvey Miller, 1995). See also J. B. Friedman, "The Architect's Compass in Creation Miniatures of the Later Middle Ages," *Traditio* 30 (1974): 419–29; Katherine H. Tachau, "God's Compass and Vana Curiositas: Scientific Study in the Old French Bible Moralisée," *Art Bulletin* 80, no. 1 (1998): 7–33. Akin to these depictions is that of Jesus using a compass to construct pools in the famous Tring Tiles, associated with the parish church at Tring, Hertfordshire, held in the British Museum.

79. Mary Carruthers, *The Experience of Beauty in the Middle Ages* (Oxford: Oxford University Press, 2013), 40.

80. On Chaucer's penchant for moving between overhead and ground-level views, see Marion Turner, *Chaucer: A European Life* (Princeton, N.J.: Princeton University Press, 2019), 217–38.

THREE. LEARNING DEVICES

1. The miniatures found a distinguished place near the start of carefully planned and executed manuscripts, all of which were likely copied in the last years of Gower's life: San Marino, California, Huntington Library, MS HM 150, fol. 13v; London, British Library, MS Cotton Tiberius A.iv, fol. 9v; University of Glasgow Library, MS Hunter 59, fol. 6v.

2. On the density and complexity of graphic page design in the period, and what Jeffrey F. Hamburger calls the "diagram paradigm in medieval art," see his "Mindmapping: The Diagram as Paradigm in Medieval Art— and Beyond," in *The Visualization of Knowledge in Medieval and Early Modern Europe*, ed. Marcia Kupfer et al. (Turnhout, Belgium: Brepols, 2020), 61–86, as well as other contributions to this volume. The idea that "diagrams were central to the making of medieval art and that much, although hardly all, of medieval art is diagrammatic," is developed at length in Hamburger, *Diagramming Devotion: Berthold of Nuremberg's Transformation of Hrabanus Maurus's Poems in Praise of the Cross* (Chicago: University of Chicago Press, 2019). See also Ayelet Even-Ezra, *Lines of Thought: Branching Diagrams and the Medieval Mind* (Chicago: University of Chicago Press, 2021), for provocative parallels.

3. James Simpson, *Sciences and the Self in Medieval Poetry: Alan of Lille's "Anticlaudianus" and John Gower's "Confessio amantis"* (Cambridge: Cambridge University Press, 1995).

4. Russell Peck, "John Gower: Reader, Editor, and Geometrician 'for Engelondes sake,'" in *John Gower: Manuscripts, Readers, Contexts*, ed. Malte Urban (Turnhout, Belgium: Brepols, 2009), 11–37; and "Gower and Science," in *Routledge Research Companion to John Gower*, ed. Ana Sáez-Hidalgo, Brian Gastle, and R. F. Yeager (New York: Routledge, 2017), 172–96.

5. Parenthetical citations are drawn from John Gower, *Confessio amantis*, in *The English Works of John Gower*, ed. G. C. Macaulay, vol. 1 (London: K. Paul, Trench, Trübner, 1900–1901).

6. The liberal arts also cast their shadow over Chaucer's *Canterbury Tales*, but they do not pool around an individual so much as get parceled out among pilgrims. According to Ann W. Astell, *Chaucer and the Universe*

of Learning (Ithaca, N.Y.: Cornell University Press, 1996), 163–78, the tales and characters of the Wife of Bath, the Clerk, and the Squire exemplify and adapt the arts of the *trivium* (rhetoric, logic, and grammar respectively).

7. Hugh of St. Victor, *The Didascalicon of Hugh of St. Victor: A Medieval Guide to the Arts,* trans. Jerome Taylor (New York: Columbia University Press, 1991), 88; for the Latin, see *Hugonus de Sancto Victore Didascalicon, De Studio Legendi: A Critical Text,* ed. Charles Henry Buttimer (Washington, D.C.: Catholic University Press, 1939), 55.

8. See Rita Copeland and Ineke Sluiter, eds., *Medieval Grammar and Rhetoric: Language Arts and Literary Theory,* AD 300–1475 (Oxford: Oxford University Press, 2009), and the essays bearing on an early modern manifestation of the phenomena in Christopher S. Celenza, ed., *Angelo Poliziano's "Lamia": Text, Translation, and Introductory Studies* (Leiden: Brill, 2010). On the emergent place of poetry in liberal arts schemes— where it retained a powerfully ambiguous role between and among other disciplines—see Mary Franklin-Brown, "Poetry's Place in Scholastic Taxonomies of Knowledge," in *Taxonomies of Knowledge: Information and Order in Medieval Manuscripts,* ed. Emily Steiner and Lynn Ransom (Philadelphia: Schoenberg Institute, 2015), 56–79.

9. Giles E. M. Gasper et al., *The Scientific Works of Robert Grosseteste,* Volume 1 (Oxford: Oxford University Press, 2019), 41.

10. See Marian Füssel, "The Conflict of the Faculties: Hierarchies, Values and Social Practices in Early Modern German Universities," in *History of Universities,* vol. 25/2, ed. Mordechai Feingold (Oxford: Oxford University Press, 2011), 80–110; and for substantive discussion of all aspects of medieval educational institutions, see William Clark, *Academic Charisma and the Origins of the Research University* (Chicago: University of Chicago Press, 2006); and Alan B. Cobban, *The Medieval English Universities: Oxford and Cambridge to c. 1500* (London: Routledge, 1988).

11. Hugh of St. Victor, *Didascalicon:* "Theorica dividitur in theologiam, mathematicam, et physicam. vel aliter, theorica dividitur in intellectibilem, intelligibilem, et naturalem. vel aliter, theorica dividitur in divinalem, in doctrinalem, et philologiam" (Buttimer, 37; Taylor, 73).

12. Hugh of St. Victor, *Didascalicon:* "Sunt enim quasi optima quaedam instrumenta et rudimenta quibus via paratur animo ad plenam philosophicae veritatis notitiam" (Buttimer, 53; Taylor, 87).

13. Hugh of St. Victor, *Didascalicon*: "Hinc trivium et quadrivium nomen accepit, eo quod his, quasi quibusdam viis, vivax animus ad secreta sophiae introeat" (Buttimer, 54; Taylor, 87).

14. Hugh of St. Victor, *Didascalicon*: "Non una tamen via omnes currunt" (Buttimer, 71; Taylor, 35).

15. Hugh of St. Victor, *Didascalicon*: "Noli multiplicare diverticula quoadusque semitas didiceris. securus discurres cum errare non timueris" (Buttimer, 57; Taylor, 90).

16. Buttimer, *Hugonus de Sancto Victore Didascalicon*, 56; Taylor, *Didascalicon of Hugh of St. Victor*, 89–90.

17. As Hugh of St. Victor writes, "*logos* means either word or reason, hence logic can be called either a linguistic or a rational science" ("dicitur enim logos sermo sive ratio, et inde logica sermocinalis sive rationalis scientia dici potest" [Buttimer, *Hugonus de Sancto Victore Didascalicon*, 21; Taylor, *Didascalicon of Hugh of St. Victor*, 59]).

18. "De ratione autem disserendi Boethius dicit quod et pars esse possit et instrumentum ad philosophiam, sicut pes, manus, lingua, oculi, etc., partes sunt corporis et instrumenta" (Buttimer, *Hugonus de Sancto Victore Didascalicon*, 45; Taylor, *Didascalicon of Hugh of St. Victor*, 80).

19. W. H. Stahl, ed. and trans., *Martianus Capella and the Seven Liberal Arts*, vol. 2, *The Marriage of Philology and Mercury*, trans. W. H. Stahl (New York: Columbia University Press, 1977), 218, 317.

20. Thierry of Chartres, *Prologue to the Heptateuch*: "Et trivium quadrivio ad generose nationis phylosophorum propaginem quasi maritali federe copulavimus" (excerpted from Édouard Jeauneau, "*Le Prologus in Eptatheucon* de Thierry de Chartres," *Mediaeval Studies* 16 [1954]: 174; the prologue is translated and discussed in Copeland and Sluiter, *Medieval Grammar and Rhetoric*, 439–43).

21. Thierry of Chartres, *Prologue to the Heptateuch*: "Nam, cum sint duo precipua phylosophandi instrumenta, intellectus eiusque interpretatio, intellectum autem quadruvium illuminet, eius vero interpretationem elegantem, rationabilem, ornatam trivium subministret" (Jeauneau, 174).

22. Thierry of Chartres, *Prologue to the Heptateuch*: "manifestum est Eptatheucon totius phylosophye unicum ac singulare esse instrumentum" (Jeauneau, 174).

23. See Deborah L. Black, *Logic and Aristotle's "Rhetoric" and "Poetics" in Medieval Arabic Philosophy* (Leiden: Brill, 1990); John W. Watt, "From

Themistius to Al-Farabi: Platonic Political Philosophy and Aristotle's *Rhetoric* in the East," *Rhetorica* 13, no. 1 (1995): 17–41. And for an English translation of the major commentaries, see *Three Arabic Treatises on Aristotle's Rhetoric: The Commentaries of Al-Farabi, Avicenna, and Averroes,* trans. Lahcen Elyazghi Ezzaher (Carbondale: Southern Illinois University Press, 2015).

24. Black, *Logic,* 181. As Black elaborates there, rhetoric aims at a broad appeal and applicability (consisting of a logic of assent) while poetics aids the powers of perception and provocation (forming a logic of apprehension).

25. Al-Farabi's *Didascalia* was translated and elaborated in Latin by Hermannus Alemannus (ca. 1256), who lists the works as belonging to the new *Organon;* see *Al-Farabi, Deux Ouvrages Inédits sur la Rhéthorique,* ed. and trans. J. Langhade and M. Grignaschi (Beirut: Dar el-Mashreq, 1971), 210; Copeland and Sluiter, *Medieval Grammar and Rhetoric,* 750. For other relevant sources that explain the aims of rhetoric—called a broader path than dialectic, and one that pertains to many practical occupations—see *Kitab al-Khataba,* in Langhade and Grignaschi, *Al-Farabi,* 56; trans. in Ezzaher, *Three Arabic Treatises.*

26. Gerard of Cremona, *Le "De scientiis Alfarabii" de Gérard de Crémone: Contribution aux problèmes de l'acculturation au XIIe siècle,* ed. Alain Galonnier (Turnhout, Belgium: Brepols, 2016); Domingo Gundisalvo, *De scientiis,* ed. Manuel Alonso Alonso (Madrid: Consejo superior de investigaciones cíentíficas, 1954).

27. Dominicus Gundissalinus, *De divisione philosophiae,* ed. L. Baur, Beiträge zur Geschichte der Philosophie des Mittelalters 4/2–3 (Münster: Aschendorff, 1903); partly trans. Copeland and Sluiter, *Medieval Grammar and Rhetoric,* 463–83.

28. Gundissalinus, *De divisione philosophiae,* trans. Copeland and Sluiter, *Medieval Grammar and Rhetoric,* 476.

29. Gundissalinus, *De divisione philosophiae,* trans. Copeland and Sluiter, *Medieval Grammar and Rhetoric,* 479.

30. Gundissalinus, *De divisione philosophiae,* trans. in Copeland and Sluiter, *Medieval Grammar and Rhetoric,* 483. Al-Farabi had earlier defined rhetoric as a "syllogistic art" in a commentary on the *Rhetoric,* on which see the opening of al-Farabi's *Kitab al-Khataba* in Langhade and Grignaschi, *Al-Farabi,* 30.

31. John of Salisbury, *Metalogicon* 1.11; for the Latin, see *Ioannis Saresberiensis, Metalogicon,* ed. J. B. Hall and K. S. B. Keats-Rohan, Corpus Christianorum Continuatio Mediaevalis 98 (Turnhout, Belgium: Brepols, 1991), 29; trans. in *The Metalogicon of John of Salisbury: A Twelfth-Century Defense of the Verbal and Logical Arts of the Trivium,* trans. Daniel D. McGarry (Berkeley: University of California Press, 1955), 33.

32. John of Salisbury, *Metalogicon* 2.12 (Hall and Keats-Rohan, 74; McGarry, 102).

33. John of Salisbury, *Metalogicon* 1.21: "Patet ex his non circa unum grammaticam occupari, sed ad omnia quae uerbo doceri possunt ut eorum capax sit, animum praeformare" (Hall and Keats-Rohan, 48; McGarry, 60).

34. See Rosemary Barton Tobin, "The Cornifician motif in John of Salisbury's *Metalogicon,*" *History of Education* 13, no. 1 (1984): 6.

35. John of Salisbury, *Metalogicon* 1.21: "Ad laudem suam referendum putant quidam nostrorum quod sine grammatica garriunt, eamque credunt inutilem et palam culpant, gloriantes se ei operam non dedisse" (Hall and Keats-Rohan, 48; McGarry, 60); for the range of the other complaints lodged against Cornificians see 1.7–9 (Hall and Keats-Rohan, 26–31).

36. Robert Kilwardby, *De ortu scientiarum,* ed. Albert G. Judy (Toronto: Pontifical Institute of Mediaeval Studies, 1976), 218.

37. Kilwardby, *De ortu scientiarum*: "Ut puteus unius tantum est per possessionem, multorum tamen per usum" (Judy, 154).

38. Kilwardby, *De ortu scientiarum*: "Et sic est omnium instrumentum et omnibus adminiculans" (Judy, 156).

39. See Steven J. Livesay, *Theology and Science in the Fourteenth Century: Three Questions on the Unity and Subalternation of the Sciences from John of Reading's Commentary on the Senteces* (Leiden: Brill, 1989), 20–53 (ch. 2).

40. Kilwardby, *De ortu scientiarum*: "Diversitas enim scientiae est ex diversitate subiecti; subalternatio ex eo quod subiectum est sub subiecto, praecipue cum descendat demonstratio a superiori in inferiorem, quae duo sese compatiuntur" (Judy, 45).

41. Kilwardby, *De ortu scientiarum*: "Ad hoc enim quod applicetur demonstratio geometrica astronomicae scientiae cadunt quaedam instrumenta media, ut praedixi, et haec instrumenta arte geometrica mechanice fiunt, ut astrolabium et instrumentum considerationis et sphaera et consimilia, et continentur sub machinativa sicut caementaria et Carpentaria et

omnino architectonica. Et ita per machinativam potest subintelligi cum quibusdam aliis etiam astronomia ratione instrumentorum ei servientium super quorum consideratione fundatur demonstratio geometrica descendens in astronomiam" (Judy, 47–48).

42. Kilwardby, *De ortu scientiarum* (Judy, 160–61).

43. Kilwardby, *De ortu scientiarum*: "Videtur ergo quod et speculativae sint practicae et practicae speculativae" (Judy, 138). See also George Odell Ovitt Jr., "The Status of the Mechanical Arts," *Viator* 14 (1983): 103.

44. Brunetto Latini, *Li Livres dou Tresor,* ed. Spurgeon Baldwin and Paul Barrette (Tempe: Arizona Center for Medieval and Renaissance Studies, 2003), 3–6; trans. Paul Barrette and Spurgeon Baldwin in *The Book of the Treasure* (New York: Garland, 1993), 1–5. Aristotle's treatment can be found in the sixth chapter of *Nicomachean Ethics,* distinguishing between episteme, phronesis, and techne.

45. Copeland and Sluiter, *Medieval Grammar and Rhetoric,* 835.

46. Christopher S. Celenza, *The Italian Renaissance and the Origin of the Humanities: An Intellectual History, 1400–1800* (Cambridge: Cambridge University Press, 2021), 117–18.

47. Poliziano, *Lamia*: "Fabulari paulisper lubet, sed ex re, ut Flaccus ait; nam fabellae, etiam quae aniles putantur, non rudimentum modo sed et instrumentum quandoque philosophiae sunt" (*Angelo Poliziano's "Lamia": Text, Translation, and Introductory Studies,* ed. Christopher S. Celenza, Brill's Studies in Intellectual History [English and Latin] [Leiden: Brill, 2010], 194–95).

48. Celenza, "Poliziano's Lamia in Context," in *Angelo Poliziano's "Lamia,"* 12.

49. Poliziano, *Lamia*: "Nostra aetas, parum perita rerum veterum, nimis brevi gyro grammaticum sepsit. At apud antiquos olim tantum auctoritatis hic ordo habuit ut censores essent et iudices scriptorum omnium soli grammatici" (Celenza, 244–45).

50. Poliziano, *Lamia*: "Grammaticorum enim sunt hae partes, ut omne scriptorum genus, poetas, historicos, oratores, philosophos, medicos, iureconsultos excutiant atque enarrent" (Celenza, 244–45).

51. Thomas Aquinas, *Expositio super librum Boethii De trinitate,* q. 5, a. 1, reply to objection 3: "Vel ideo hae inter ceteras scientias artes dicuntur, quia non solum habent cognitionem, sed opus aliquod, quod est immediate ispius rationis, ut constructionem, syllogismum vel orationem formare, numerare, mensurare, melodias formare et cursus siderum computare"

(ed. Bruno Decker [Leiden: Brill, 1959], 168; parts of the commentary are available in English in Aquinas, *The Division and Methods of the Sciences: Questions V and VI of His Commentary on the "De trinitate" of Boethius,* trans. Armand Maurer, 4th rev. ed. [Toronto: Pontifical Institute of Mediaeval Studies, 1986], 18). Thomas goes on to distinguish sharply between the mechanical and liberal arts, since higher sciences have no other products than knowledge. Others foregrounded in the present book recognized a more fundamental interdependency among the various arts and sciences.

52. Cobban, *Medieval English Universities,* 161.

53. For a survey of views, see Ovitt, "Status," and especially Elspeth Whitney, *Paradise Restored: The Mechanical Arts from Antiquity through the Thirteenth Century* (Philadelphia: American Philosophical Society, 1990).

54. See Gerard of Cremona, *Le "De scientiis Alfarabii" de Gérard de Crémone,* 256–60; Gundissalinus, *De scientiis,* 108–12, and *De divisione philosophiae,* 122–24. In their accounts, mathematical theory conveys information about intelligible quantities (lines, surfaces, numbers, etc.) apart from natural bodies; mechanical devices present information to the senses by endowing receptive matter with the capacity to apply knowledge. It is by means of some artful invention ("ad inveniendum per artem") that knowledge becomes evident and useful in this way.

55. Kilwardby, *De ortu scientiarum* (Judy, 129–33); Ovitt Jr., "Status," 101–5.

56. Hugh of St. Victor, *Didascalicon:* "Mechanica septem scientias continet: lanificium, armaturam, navigationem, agriculturam, venationem, medicinam, theatricam. . . . ad similitudinem quidem trivii et quadrivii, quia trivium de vocibus quae extrinsecus sunt et quadrivium de intellectibus qui intrinsecus concepti sunt pertractat" (Buttimer, 38–39; Taylor, 74). For an assessment of Hugh's mechanics in relation to the history of technology, see Eric Schatzberg, *Technology: Critical History of a Concept* (Chicago: University of Chicago Press, 2018), 32–36.

57. Hugh of St. Victor, *Didascalicon:* "Hae sunt septem ancillae quas Mercurius a Philologia in dotem accepit, quia nimirum eloquentiae, cui iuncta fuerit sapientia, omnis humana actio servit" (Buttimer, 39; Taylor, 75). On John the Scot, see Whitney, *Paradise Restored,* 70.

58. On the distinction between learned and living languages, see Karla Mallette, *Lives of the Great Languages: Arabic and Latin in the Medieval Mediterranean* (Chicago: University of Chicago Press, 2021), 10.

59. A proverbial notion of writing as plowing goes back to Plato. For example, Egbert of Liege wrote: "With this pen and stylus we furrow the field's acres" (*The Well-Laden Ship,* trans. Robert G. Babcock [Cambridge, Mass.: Harvard University Press, 2013], 121). On plowing as a manuscript metaphor, see Julia Holloway Bolton, *The Pilgrim and the Book: A Study of Dante, Langland and Chaucer,* 3rd ed., American University Studies (New York: Peter Lang, 1993), 197.

60. Lisa H. Cooper, "Agronomy and Affect in Duke Humfrey's *On Husbandrie,*" *Speculum* 95, no. 1 (2020): 62.

61. Lisa H. Cooper, *Artisans and Narrative Craft in Late Medieval England* (Cambridge: Cambridge University Press, 2011), 25–29, observes that tool lists present them, in Heidegger's terms, as "ready-to-hand."

62. Patricia Clare Ingham, *The Medieval New: Ambivalence in an Age of Innovation* (Philadelphia: University of Pennsylvania Press, 2015).

63. Isidore of Seville, *Etymologies* 10.1.122, in *The Etymologies of Isidore of Seville,* ed. and trans. Stephen A. Barney, W. J. Lewis, J. A. Beach, and Oliver Berghof (Cambridge: Cambridge University Press, 2006).

64. Cassiodorus, *Institutiones* 2.4, ed. R. A. B. Mynors (Oxford: Clarendon, 1961).

CONCLUSION

1. Martin Heidegger, *Being and Time,* trans. John Macquarrie and Edward Robinson (Oxford: Blackwell, 1962), 97–107.

2. Heidegger, *Being and Time,* 97. Later, in *The Question Concerning Technology and Other Essays,* trans. William Lovitt (New York: Garland, 1977), 5–7, Heidegger elaborates on the failings of an "instrumental definition of technology" caught up in means, causes, and consequences.

3. Gilbert Simondon, *The Mode of Existence of Technical Objects,* trans. Cecile Malaspina and John Rogove (Minneapolis, Minn.: Univocal, 2017), 16.

4. Bernard Stiegler, *Technics and Time, 1: The Fault of Epimetheus,* trans. Richard Beardsworth and George Collins (Stanford, Calif.: Stanford University Press, 1994), ix.

5. Stiegler, *Technics and Time, 1,* 206.

6. For some critiques, see Debra Benita Shaw, *Technoculture: The Key Concepts* (New York: Routledge, 2008), among other feminist and queer studies of science, technology, and ecology that have helped to identify the reigning paradigm behind such worries. See, for example,

Donna J. Haraway, "A Cyborg Manifesto: Science, Technology, and Socialist Feminism in the Late Twentieth Century," in *Simians, Cyborgs and Women: The Reinvention of Nature* (New York: Routledge, 1991), 149–81; Elizabeth Grosz, *Time Travels: Feminism, Nature, Power* (Crows Nest, Australia: Allen & Unwin, 2005); Stacy Alaimo, *Bodily Natures: Science, Environment, and the Material Self* (Bloomington: Indiana University Press, 2010); Jane Bennett, *Vibrant Matter: A Political Ecology of Things* (Durham, N.C.: Duke University Press, 2010); Katherine Behar, ed., *Object-Oriented Feminism* (Minneapolis: University of Minnesota Press, 2016).

7. Compare the concluding manifesto to Eric Schatzberg, *Technology: Critical History of a Concept* (Chicago: University of Chicago Press, 2018), 235, in which he expresses a hope that scholars will "rescue technology from determinists." For Schatzberg, the problem stems from the reduction of technology to instrumentalism (232: "Determinism, whether expressed by enthusiasts or by pessimists, is tightly linked to an instrumental conception of technology that divorces it from culture"), whereas I would say that determinism originates in a narrow, technologized conception of instrumentality itself. Nonetheless, a renewed understanding of technology, of the sort offered by Schatzberg, should make common cause with any attempt to reconstruct or rehabilitate instrumentality. It is just that he has not extended the main insight far enough, to see instrumentalism as another element of culture and the arts.

8. See Jesse Oak Taylor, "Globalize," in *Veer Ecology: A Companion for Environmental Thinking,* ed. Jeffrey Jerome Cohen and Lowell Duckert (Minneapolis: University of Minnesota Press, 2017), 34, on how the physical globe is a "world-making model" that made globalization possible.

9. Antonio Gramsci, *Selections from the Prison Notebooks,* ed. Quintin Hoare and Geoffrey Nowell Smith (New York: International, 1971), 143.

10. Darrow Schecter, *The Critique of Instrumental Reason from Weber to Habermas* (New York: Continuum, 2010), 78.

11. Charles Taylor, *Sources of the Self: The Making of the Modern Identity* (Cambridge, Mass.: Harvard University Press, 1989), and *A Secular Age* (Cambridge, Mass.: Belknap, 2007).

12. Stanley Fish, *Think Again: Contrarian Reflections on Life, Culture, Politics, Religion, Law, and Education* (Princeton, N.J.: Princeton University Press, 2015), xviii.

13. Gramsci argued for a publicly funded system that would look after the "humanistic formation" of the young, leading them instrumentally through a set curriculum (*Selections,* 170–71).

14. Louis Menand, *The Marketplace of Ideas: Reform and Resistance in the American University* (New York: Norton, 2010), 53.

15. This is the thrust of William Clark, *Academic Charisma and the Origins of the Research University* (Chicago: University of Chicago Press, 2006).

16. Alan B. Cobban, *The Medieval English Universities: Oxford and Cambridge to c. 1500* (London: Routledge, 1988), 10.

17. Frank Donoghue, *The Last Professors: The Twilight of the Humanities in the Corporate University* (New York: Fordham University Press, 2008), 22.

18. Eric Hayot, Humanist Reason: *A History. An Argument. A Plan* (New York: Columbia University Press, 2021), 18. Later in the book, he identifies a series of epistemological principles that have been developed within the humanities, including: "all Human activity is context-embedded, but not context-determined"; "human social life is not flat"; "scales are complex, overlapping, and porous"; "fuzziness, ambiguity, and contradiction are socially functional."

19. John Guillory, *Professing Criticism: Essays on the Organization of Literary Study* (Chicago: University of Chicago Press, 2022), 351–52. He lists five rationales in total for the study of literature: linguistic/cognitive, moral/ judicial, national/cultural, aesthetic/critical, and epistemic/disciplinary.

20. Rita Felski, *Hooked: Art and Attachment* (Chicago: University of Chicago Press, 2020), 130; and see her much earlier *Uses of Literature* (Oxford: Blackwell, 2008).

21. Paulo Freire, *Pedagogy of the Oppressed: 50th Anniversary Edition,* trans. Myra Bergman Ramos (New York: Bloomsbury, 2000), 48, 180. At various points in the book, Freire contrasts *instrumentos* of liberation, rehumanization, and cultural revolution with those wielded against the oppressed.

22. bell hooks, *Teaching to Transgress: Education as the Practice of Freedom* (New York: Routledge, 1994), 12, 65.

23. Sarah Ahmed, *What's the Use?* (Durham, N.C.: Duke University Press, 2019), 222.

24. University of Victoria, *Indigenous Plan, 2017–2022,* uvic.ca/assets2012 /docs/indigenous-plan.pdf, accessed December 2021.

INDEX

abstractions, 30, 48, 50, 59, 61, 64, 80, 124n64; astrophysical, 69; concrete, 112n24

Abu Ma'Shar, 25

Accursius de Parma, 101n2

action, 3, 41, 51, 99, 100; communicative, 84; cultural, 98; intellection and, 52; thought and, 17, 57, 68

Actor Network Theory (ANT), 107n43, 116n59

Adam of Petit Pont, 87

Adelard of Bath, 21, 25, 34

Adorno, Theodor W., 94

aestheticism, 17, 96

agency, 4, 12, 42, 97, 99; causal, 11; individual, 9, 42; intellectual, 3

Ahmed, Sarah, 98

al-Andalus, 25

Albertus Magnus, 38

Alexander, 70

al-Farabi, 77, 130n25, 130n30; *scientia ingeniorum* and, 85

Alfonso, King, 49

alidade, 19, 26

Alkabucius, 25

al-Kwarizmi, 25

Allen, Valerie, 104n14

Almagest (Ptolemy), 25, 55 (fig.), 58, 64, 66

almucantar, 26

Alpha Aquilae, 117n3

Alpha Lyrae, 117n3

Altair, 45, 117n3

altimetric devices, 5, 52

al umm, 34, 36

ankabut, 37

ANT. *See* Actor Network Theory

anti-instrumentality, 10, 17, 42, 97

Apollo, 43

archaeology, media, 12, 13

archer portrait, 71 (fig.)

archery, 61, 62, 64

arcus, 61, 70

Aristarchus, 88

Aristotelians, 78, 82

Aristotle, 6, 8, 38, 49, 70, 73, 76, 77, 80, 81, 82, 132n44

arithmetic, 16, 27, 40, 72, 80

armillary spheres, 2, 25, 65

Arnold, Matthew, 9

Ars Poetica (Horace), 104n14

artes liberales, 72, 89

artes mechanicae, 41, 76, 83, 84–89

arts, 20, 74, 81, 84, 86, 96, 97, 99;
amatory, 89; contemporary,
98; curricula, 95; historians of,
48; instrumental, 76; literary, 7;
mathematical, 72; medieval, 73,
112n26; religious, 51; sciences and,
72, 86, 133n51; synthetic vision of,
81; verbal, 73. *See also* language
arts; liberal arts; mechanical arts

Astrolabe (son), 21

astrolabe plate, diagram of, 39 (fig.)

astrolabes, 2, 3, 4, 5, 21, 25, 45, 59;
boundaries and, 31; Chaucer,
31; contemporary timekeeping
devices and, 32; coordinating
figures and, 32–33; early English,
22 (fig.); emergence of, 19–20;
equatorial, 30; front of, 35
(fig.); geological time and, 36;
hemispheric vault and, 29; as
multimodal technology, 43;
planispheric, 19, 54; pointers on,
45; utility of, 20

Astrolabium (Māshā'alāh), 25

astronomy, 1, 2, 15, 16, 24, 25, 26,
43, 46, 55, 56, 70, 72, 76, 80, 93;
mathematical, 58; practical,
22, 30, 54, 58; recreational, 49;
spherical, 53, 61; technical, 67;
theory/practice of, 59

astrophysical phenomena, 2, 28–29

azimuth, 26, 37

Bacon, Francis, 3–4

Bacon, Roger, 8, 21, 40, 54

Barad, Karen, 12, 107n43

Barber, Peter, 49

Bartholomaeus Anglicus, 7–8, 38,
102–3n8

Bede, 103n11

Bennett, Jane, 36

bestiaries, 38

Bible moralisée, 67

bismillah, 25

Boethius, 7

Bogen, Steffen, 49

Bonaventure, 79

Book of the Treasure (Brunetto), 81

Borrelli, Arianna, 23, 28, 109n10

Breviculum (Le Myésier), 121n38

British Library, MS Harley 334, 67

British Library, MS Harley 3647, 25

Brunetto Latini, 79, 81

Buridan, John, 125n69

Cadmus, 88

calculations, 5, 19, 25, 54, 62, 65, 76

calendars, 27, 30, 48

Canadian Truth and Reconciliation
Commission, 99

Canterbury Tales, The (Chaucer), 20,
127n6

Capella, Martianus, 76, 85

capitalism, 91, 99; global, 93, 94;
racial, 11

Carmentis, 88

cartography, 11, 93

Cassiodorus, 89

Categories (Aristotle), 6

Cato, 87

causation, 6, 98

Celenza, Christopher, 83

celestial bodies, 56, 72

Châtelet, Gilles, 53

Chaucer, Geoffrey, 3, 15, 20, 21, 30, 40, 43, 45, 64, 67; almucantars and, 29; *Astrolabe* and, 5, 19, 22–23, 25–26, 28, 31, 33, 34, 108n2, 114n40, 117n3; language of, 23; scientific inquiry and, 44; technical apparatus of, 24, 28; theory/practice of, 23; *translatio studii* and, 27; xenoglossia and, 27; zodiac band and, 33

chords, 37, 46, 55, 56, 56 (fig.), 57 (fig.), 61, 62, 67; arcs of, 59; diagram of, 64; double, 59; formal equalities and, 64; lengths, 54; Ptolemaic, 57; theory, 54; triangles and, 58

chord tables, 55 (fig.)

Chronica maiora (Paris), 50, 52

Cicero, 7, 88

Clark, Andy, 40

Clark, William, 10

climate change, 11, 99

Cobban, Alan B., 97

cognition, 16, 20, 23, 41, 46, 48, 58; modes of, 98; multiscalar models and, 28–31

colonialism, 93, 99

combinatorics, 51, 52

Commedia (Dante), 67

commerce, 85, 91, 105n28

communication, 6, 8, 59, 86; strategies, 84; technical, 48; transcultural, 89; visual, 10

conarachne, 37

conceptual system, 51, 62

Confessio amantis (Gower), 70, 86, 88

Connolly, Daniel K., 50–51, 52

Connor, Steven, 12

Cooper, Lisa, 87

Copeland, Rita, 81

Cornificians, 79

cosmimetria, 55–56

cosmology, 30; antique, 24; vernacular, 27

cosmopolitics, 107n43

cosmos, 22, 28, 31, 41, 64, 67, 72

Creator as Geometer, 67, 126n78

criticism, 10, 17; cultural, 5, 82

criticus, 83, 84

Cucker, Felipe, 121n34

cultural criticism, 5, 82

culture, 17; aesthetic, 93; diagram, 49; literary, 20, 86; nature and, 12, 13, 23, 99; scholastic, 49; scientific, 14, 24, 25

cylinder, 2, 101n2, 126n70

Dante, 67, 81

data, 109n10; astronomical, 15; auxiliary, 19

De artibus liberalibus (Grosseteste), 79

De compositione et operacione astrolabii (Māshā'allāh ibn Atharī), 24

De divisione philosophiae (Gundissalinus), 77

De Landa, Manuel, 36

Delano-Smith, Catherine, 49

Deleuze, Gilles, 53, 115n43

demonstrations, 54; geometrical, 80; rational, 80

De nominibus utensilium (Nequam), 87

De nuptiis Philologiae et Mercurii
(Capella), 76
De ortu scientarium (Kilwardby), 79
De reductione artium ad theologian
(Bonaventure), 79
De sphaera (Sacrobosco), 25
determinism, 135n7; technological,
93, 94
De utensilibus (Adam of Petit Point),
87
diagrammata, virtues of, 54
diagrams, 54; chord, 46, 55, 56, 56
(fig.), 57, 64; drawing, 47–48;
geometrical, 46; kinematics
of, 53; philosophical tree, 49;
prevalence of, 73; rotational, 53;
technical, 52; visual, 23, 45, 59,
119n17
Dictionarius (John of Garland), 87
Didascalia (Al-Farabi), 130n25
Didascalicon (Hugh), 74, 88
digital age, 9, 94
distinctiones, 47
divisio scientiae, 70, 83
divisio textus, 47
documentum, 46
Donatus, 88
Donoghue, Frank, 97

Eagle star Altair, 45, 117n2
ecology, 96; feminist/queer studies
of, 134n6; translation/translin-
gual, 24–28
education, 91; civic, 79; higher, 83,
99; literary, 83; models of, 97; theory, 73. *See also* learning; liberal
education

Egbert of Liege, 134n59
embodiment, 20, 28, 36, 38, 40, 52, 53
empirical phenomena, 19, 24
encyclopedias, 38, 47, 70
engyn, 37
Enumeration of the Sciences (Al-
Farabi), 77
epistēmē, 81; *technē* and, 10
epistemic objects, 16, 43, 69
Equatorie of the Planetis, 25, 108n2
equatorium, 4, 21, 28, 46; described,
102n3; making, 47; planetary, 25;
treatises for, 2, 102n3
equipmentality, 92
errare, 75
Etymologies (Isidore of Seville), 5,
6, 47
Euclid, 25, 29, 54, 57, 121n53
Eudoxus, 37
Even-Ezra, Ayelet, 49
exchange: cross-cultural, 15, 25, 26,
99; transnational, 14, 15

fabella, 82
Falk, Seb: equatorium and, 110n16
Felski, Rita, 98
Ferraris, Maurizio, 126n70
Ficino, Marsilio, 83
fictus, 31, 61
figmenta rationis, 61
figures: codex-bound, 52; coordinating, 32–33; diagrammatic, 16, 50;
exemplary, 61; flat, 46–49, 55, 64,
67; in flight, 65, 67–68; geometrical, 59; graphic, 46; scientific, 61;
transdisciplinary poetic, 72
fingere, 31, 61

Fish, Stanley, 94, 95
flight: extraterrestrial, 65; inhuman, 65, 67–68
fold-outs, 50
formae, 61
formalism, 12, 51, 70
Frankfurt School, 9, 94
Freire, Paulo, 98, 136n21

geography, 22, 46, 49, 54
Geography (Ptolemy), 50
geometers, 70, 72
Geometria (Gerbert), 58, 62
geometry, 16, 22, 46, 52, 64, 70, 72, 76; abstract, 29; Euclidean, 53; mathematical, 53–59, 80, 124n64; medieval, 53–54; plane, 29, 55–56; practical, 62; spherical, 24; universal, 30
Gerard of Cremona, 25, 66, 77, 80, 85; chord tables and, 55 (fig.)
Gerbert of Aurillac, 25, 57, 62, 63; triangles and, 58
Giardino, Valeria, 48, 51
globalization, 93, 135n8; neoliberal, 11
globes, 21, 28; manual for, 46; spherical, 65
Goldie, Matthew Boyd, 50, 52
Gower, John, 69, 74, 81, 86, 88, 89; curricular models and, 73; formulation by, 82; knowledge practices and, 72; labor and, 87; moral geometer and, 70
grammar, 16, 73, 76, 78, 84; logic and, 81–82
grammaticus, 83, 84
Gramsci, Antonio, 94, 95, 136n13

Grant, Edward, 20–21
graphic field, 46, 51, 118n8
graphism, 64
Grosseteste, Robert, 7, 79, 80
Grosz, Elizabeth, 12
gubernator, 32
Gundissalinus, Dominicus (Domingo Gundsalvo), 74, 77, 78, 80, 85, 86

Hall of Fame (Chaucer), 64
Hamburger, Jeffrey F., 48, 51, 127n2
hand, 4, 8–9, 31, 47, 86, 87, 104n23; *digitus*, 9; interdigitation, 9; handbooks and, 8; instrument of instrument, 8; *manus nuda*, 4
Hankins, Thomas, 13
Harvey, E. Ruth, 116n55
Hayot, Eric, 96
Heidegger, Martin, 92, 134n2, 134n61
Heptateuch (Thierry), 76
Hermannus Alemannus, 130n25
Herodotus, 88
Hipparchus of Nicaea, Ptolemy and, 54
historiated initial, 66 (fig.)
homo ludens, 120n26
hooks, bell, 98
Horace, 104n14
horizons, 25, 28, 67; cultural, 13, 61; transcultural, 15; transhemispheric, 15, 24
Horobin, Simon, 20
House of Fame (Chaucer), 43, 45, 67
Hugh of St. Victor, 6, 56, 72, 74–75, 77, 80, 88; art and, 75; on *logos*, 129n17; triangles and, 58

Huizinga, Johan, 120n26

humanism, 83, 136n13; Florentine, 82; literary, 10, 95

humanities: digital, 12, 106n30; instrumental defense of, 94

hybrids, technoscientific, 31–34, 36–38, 40–42

hypotheticals, 16, 21, 30

imagination, 61; acts of, 77; scientific, 45, 46, 50, 65; technical, 34, 70, 93

information capture, 20

information systems, 11, 47, 91

ingenium, 78

Ingham, Patricia, 87

Ingold, Tim, 40

inscriptions, 12, 15, 41, 43, 44, 46, 48

instrumenta, 3, 5, 6, 7, 8, 13, 16, 38, 47, 73, 75, 76, 81, 83, 103n11, 119n16; term, 5, 78

instrumental rationality, 9, 98, 105n25

instrumenta philosophiae, 77, 82

instrumenta scribae, 92

instrumenta sensum, 40

instrumentation, 11, 13, 99; astronomical, 67; geometrical, 67

instruments, 48, 75, 78, 86, 94, 119n14; capacity for work and, 16; contrived, 8; critical, 12; earliest recorded use of, 5; intellectual, 54, 74; mathematical, 85, 105n28; natural, 86; philosophy, 77, 82; scientific, 11, 20, 105n28; technical, 4; verbal, 79

"Instruments of Good Works" (*instrumenta bonorum operum*), 6–7

intellection, 52, 76

interdependency, 79, 133n51

interdisciplinary, 14, 89

interpretation, 11, 76, 94

inventio, 86, 87, 88

Isidore of Seville, 5, 6, 47, 92

John of Garland, 87

John of Harlebeke, 1, 2, 101n2

John of Lignières, 2, 15, 102n3

John of Salisbury, 6, 61, 74, 78, 79, 103n11

kataskopos, 67, 70, 72

Kaye, Joel, 125n69

Kilwardby, Robert, 6, 74, 80, 81, 85

knowledge, 96; acquiring/transmitting, 44; astronomical, 28; craft, 86; dissemination of, 89, 99; instrument, 9; intensification of, 23; mobilization of, 3, 5; object of, 3, 43; organization of, 46; portable, 24; practical, 70; theoretical, 70; transcultural, 14

knowledge practices, 6, 41, 69, 72; international, 15; premodern, 17

knowledge production, 3, 23, 36, 45, 86; conditions of, 84

Kupfer, Marcia, 50

labor: academic, 10, 97; casualized, 97; exploitation of, 93, 94

lamias, 82

language: cosmopolitan, 15; interventions of, 73; literature and, 75–76; natural, 51; status of, 79; studying, 78; visual, 51

language arts, 75, 77, 78, 83, 84

Larkin, Jill, 51

latitudes, 26, 28

Latour, Bruno, 107n43

law, 6, 96; social disorder and, 99

learning, 84; branches of, 81; debasement of, 91; grammatical, 79; higher, 6, 76, 79, 85. *See also* education

Le Myésier, Thomas, 121n38

Lerer, Seth, 114n40

Lewis, 20, 34, 44

liber, 89

liber aeneus, 45

liberal arts, 10, 16, 24, 41, 73, 74, 76, 83, 85, 86, 89, 91, 97, 127n6, 133n51; medieval, 67; poetry and, 128n8; treatises on, 84

liberal education, 5, 14, 70, 79, 84, 96, 98, 99; constructive, 97. *See also* education

L'image du monde, 55, 56, 67

line segments, 57 (fig.)

literary training, 78

literary work, use-enjoyment in, 104

literature, 82; imaginative, 67; invention of, 73, 84–89; language and, 75–76; mechanical arts and, 84–89; study of, 95

literatus, 83, 84

Ljungberg, Christina, 48

logic, 16; Aristotelian, 76; grammar and, 81–82; status of, 79; studying, 78; visual, 70

logique, 81

logos, 12, 73, 75, 76, 77, 79, 82, 85, 86, 129n17; concept of, 84

Lull, Ramon, 51

Macbeth, Danielle, 53

machina mundi, 30

Mallette, Karla, 86

manuals, 8, 21, 28, 46, 62, 85, 91; instrumental, 118n7

manuscript production, 11, 46–47, 89

manus nuda, 4

mappae mundi, 50

maps, 48; flat, 65; geographical, 49; T-O, 50, 70; zonal, 49

Māshā'allāh ibn Atharī, 15, 24, 25, 27, 44

mater, 26, 34, 36. *See also* womb

mathematics, 31, 46, 54, 62, 64, 75, 76; models, 28, 126n70; notations, 27, 53, 122n48; signs in, 53

McCluskey, Stephen, 24

Mead, Jenna, 36, 111n18

measurement, 56, 84, 126n70; devices, 16; instruments of, 52

mechanical arts, 41, 76, 83; invention of literature and, 84–89; philology and, 85

mechanisms, 2, 15, 45; digital, 12; multipurpose, 19; physical, 59

media, 13, 14, 15, 17, 97; communicative, 89; flat, 46–49, 55, 64, 67; graphic, 41; informational, 67; inscriptive, 43; substrates, 12; technology, 3, 11, 12, 36, 69, 85

media studies, 96, 106n30

mediation, 5, 11, 94

medicine, 6, 85

Menand, Louis, 96

Mercury, Philology and, 76, 85

meridian, 28

Merleau-Ponty, Maurice, 53

Metalogicon (John), 78

metaphors, 31, 40, 41, 62, 114–15n42; animal, 23; matrixial, 36; mercurial, 23–24; operative, 13

metaphysics, 83

model dependence, metaphorical transference and, 59, 61–62, 64–65

models, 5; artificial, 59; curricular, 73; graphic, 58; inventive, 21; scientific, 59

modernity, 68, 95; instrumentalizing, 91; postindustrial, 93; technocratic, 93

multiscalar models, cognition and, 28–31

music, 6, 16, 48, 72, 104n23, 105n28

nadir, 26

nature: culture and, 12, 13, 23, 99; technics of, 116n60

navicula, 28, 32, 59, 113n33, 119n16; manual for, 46

Navicula de Venetiis, 33 (fig.)

Nequam, Alexander, 87

networks: promiscuous/monogamous, 114n37; text, 14, 15

Newe Theorik of Planetis, The, 25

Nichols, Stephen G., 22

Nicomachean Ethics (Aristotle), 132n44

Novaes, Catarina Dutilh, 12

Novum Organum (Bacon), 3–4

numerals, 23, 26, 84; defense of, 12; elementary, 27; Hindu-Arabic, 14, 53; Roman, 27

On the Mode of Existence of Technical Objects (Simondon), 92

On the Properties of Things (Bartholomaeus), 7–8

On the Use of the Astrolabe (Abu Ma'Shar), 25

Opus agriculturae (Palladius), 87

orbis terrarum, 50

Oresme, Nicole, 61, 125n66, 125n69

Organon (Aristotle), 6, 49, 73, 77, 78, 82; expansion of, 76

organopoiia, 16, 21, 34

Ovid, 38

Palladius, 87

Parikka, Jussi, 36, 116n60

Paris, Matthew, 50, 52

paternalism, intellectual/national, 114n40

Peterhouse, Cambridge, MS 75.1, 25

philology, 75, 76, 83, 85

Philology, Mercury and, 76, 85

Philo of Alexandria, 38

philosophiae instrumenta, 16, 41, 67, 69, 73–84

philosophy: Aristotelian, 72, 78; branches of, 70; continental, 53; Farabian, 77; instruments of, 6, 41, 76, 77–78, 97; natural, 21, 72; poetry and, 69, 72; rudiments of, 82; study of, 95

physics, 43, 83

Pickering, Andrew, 107n43

Pico della Mirandola, 83

planet, 26

planimetria, 55–56

Planisphaerium (Ptolemy), 37

Plantagenet, Henry, 34

Plato, 134n59

play-space, 49

Pliny, 38

plumb bobs, 32, 38, 59

plumbline, 21, 32

poetics, 6, 62, 73, 78, 84

Poetics (Aristotle), 77

poetry, 77; liberal arts and, 128n8; philosophy and, 69, 72

Poliziano, Angelo, 74, 82, 83

Porphyrean trees, 53

possessio, 80; *usus*, 80, 82

Posterior Analytics (Aristotle), 80

Practical Geometry (Hugh), 58

practique, 70, 72, 81, 86

Price, Derek de Solla, 21

Prior Analytics (Aristotle), 82

prosthetics, premodern, 8, 12, 13, 41

Ptolemy, 21, 24–25, 27, 37, 44, 55, 58, 64, 65, 66; Hipparchus and, 54; latitudes/longitudes and, 49; maps and, 50; *organopoiia* of, 16

Pythagorean theorem, 53, 62; demonstrating, 63 (fig.)

quadrants, 2, 101n2

Quadripartitum (Richard), 58, 60 (fig.)

quadrivium, 16, 72, 73, 75, 76, 85

Quia nobilissima (John of Lignières), 2

quietism, 95

racism, environmental, 93

rationalization, 50, 94, 98

reasoning: geometrical, 61; mathematical, 53; practical, 17

rectangulus, 28, 59, 64

religion, 89, 99

reproductivity, 38

rhetoric, 6, 16, 24, 73, 76, 77, 78, 79, 84, 89

Rhetoric (Aristotle), 77, 130n30

rhetorique, 70, 72, 81, 86

Richard of Wallingford, 58, 59, 61, 64; prologue by, 60 (fig.)

root and model (*radix et exemplar*), 2

Rothschild Canticles, 65, 67

Rotman, Brian, 12, 52, 53, 125–26n70

Sacrobosco, John, 24, 25, 30, 44, 65

sagitta, 59, 61, 70

saphea, 2

Sawday, Jonathan, 108n4

scales, 29, 32, 36, 47, 59, 93

Schatzberg, Eric, 105n26, 135n7

science, 11, 37, 80, 96; arts and, 72, 86; astrophysical, 45, 65; feminist, 12, 134n6; history of, 14, 48; medieval, 14, 124n62; practical, 59, 85; specialized instruments of, 70; vernacularization of, 24, 26

scientia ingeniorum, 21, 85

scientia lingue, 77

"Scientific Babel," 14

scientific imagination, 20–21, 124n62; medieval, 46, 65

scientific investigation, 1, 13, 44, 58

Seaman, Myra, 104n12

Seneca, 7, 38

sensorimotor activity, 52, 54

Silverman, Robert, 13

Simon, Herbert, 51

Simondon, Gilbert, 92, 93

sines/cosines, 58

sinus rectus, versus, and *duplatus,* 58–59

Skilled Practice Involves Developmentally Embodied Responsiveness (SPIDER), 40, 116n49, 116n59

Sluiter, Ineke, 81

social change, 11, 96, 100

Somnium Scipionis, 67

space, 58; miniaturizing, 50; time and, 53

speche, 82, 86

sphera solida, 1–2

spider, 38, 40; arachne, 37; arachnoid, 40; aranea, 37; thread metaphors and, 37

SPIDER. *See* Skilled Practice Involves Developmentally Embodied Responsiveness

spiritual contemplation, 7, 51

Stadolink, Joe, 114n40

star pointers, 33

Steiner, Emily, 47

STEM, 14, 99

Stengers, Isabelle, 107n43

stereographic projection, 28–29

Stiegler, Bernard, 12, 93

strings, 61

subalternation, 80, 82, 85

surveillance: digital, 94; technological, 93

syllogisms, 77, 78, 79, 81, 82, 84

Tabulae (Toledan), 25

Taylor, Charles, 94, 95

technē, 12, 98; *epistēmē* and, 10

technical objects, 8, 13, 33, 61

technicity, 21, 91, 92–93

Technics and Time (Stiegler), 93

techniques, 11, 21, 92

technocracy, 91

technocultures, 5, 13

technological progress, 94, 113n33

technology, 2, 37, 89, 91, 92, 94; assistive, 11, 42; cognitive, 12, 70; concept of, 105n26; enduring, 93; feminist, 134n6; history of, 14, 48; information, 47; instrumental definition of, 134n2, 135n7; language of, 87; media, 3, 11, 12, 36, 69, 85; medieval, 12, 93; theories of, 11; troping of, 34

technopoesis, 21

Tertullian, 103n11

textuality, 2, 37, 41, 45

Thābit ibn Qurra, 25

Theorica planetarum (Gerard of Cremona), 25

theorique, 70, 81, 86

Thierry of Chartres, 6, 74, 76, 129n20

Thomas Aquinas, 54, 61, 84, 133n51

time: miniaturizing, 50; space and, 53

tools, 11, 13, 92; digital, 55; heuristic, 74; mechanical, 68; segregation of, 93; of thought, 48, 73

torquetum, 2, 101n2

Tractatus de sphaera solida (Sacrobosco), 1, 24, 65

transactionalism, 79, 91

translatio, 23, 27, 28, 30, 34

translation, 15, 23; theory of, 26; transmission, 41

Treatise on the Astrolabe (Chaucer), 3, 5, 21, 22–23, 25–26, 28, 29, 32, 33, 40; described, 19; Lewis and, 20,

34, 44; as training manual, 20; translingual ecology and, 27
Treharne, Elaine, 47
triangles, 64; chords and, 58; equilateral, 54, 57; mother of all figures, 58; right, 58
Tring Tiles, 126n78
trivium, 16, 72, 73, 75, 76, 79, 85
Troilus and Criseyde (Chaucer), 20

umbra versa, 26
universitas, 96
Usk, Thomas, 26

Varro, 87
vernacular, 4, 8, 25, 26, 27, 70, 96
viae, 16; paths to wisdom, 75
visualization, 40, 41, 46, 47, 48, 120n22

Vitruvius, 37
volvelles, 45, 52
von Uexküll, Jakob, 38
Vox clamantis (Gower), 69

Wakelin, Daniel, 112n24
Waters, Clarie M., 105n29
Weber, Max, 9, 94
Westwyk, John, 110n16
What's the Use? (Ahmed), 98
womb, 37, 40, 114n37

xenoglossia, 27

zenith, 26, 28, 29
zij, 37
zodiac, 2, 26, 30, 32, 33, 49

J. ALLAN MITCHELL is professor of English and director of medieval studies at the University of Victoria. He is author of *Becoming Human: The Matter of the Medieval Child* (Minnesota, 2014), *Ethics and Eventfulness in Middle English Literature*, and *Ethics and Exemplary Narrative in Chaucer and Gower*.